T0357377

Transforming the Construction Industry with Blockchain: Enhancing Efficiency, Transparency, and Collaboration

Transforming the Construction Industry with Blockchain

Enhancing Efficiency, Transparency, and Collaboration

James Harty

KEA, Copenhagen School of Design and Technology
Copenhagen, Denmark

Published by John Wiley & Sons, Inc., Hoboken, New Jersey.
Published simultaneously in Canada.

Library of Congress Cataloging-in-Publication Data applied for

Hardback ISBN: 9781394216383

Cover Design: Wiley
Cover Image(s): © Nikolay Pandev/Getty Images

Set in 10.5/13pts ChapparalPro Regular by Straive, Pondicherry, India

SKY10099015_022525

To my beloved Lene

Contents

Preface

The construction industry is in a state of flux. Some would say that is an understatement. Some would say we are building too much. Some would say we need to recycle and transform what we have. Most would agree something has to change. If the industry was a country, it would be the third worse in the world after China and the United States of America, replacing India for carbon-dioxide embodiment.

Moreover, than efficiencies, better methods are needed to bring transparency to the table so that we can reward better practices, which in turn will lead to better collaborations. Digitalisation too is at way too many differing levels, depending on where you are coming from and how you engage as a stakeholder across the whole spectrum of the industry. It is also unfolding in multiple ways, bringing a plethora of new techniques, reflecting and contradicting each other, in equal measure.

Whatever the situation, Thom Mayne of the American Institute of Architects in Las Vegas said:

> *"It's about survival. If you want to survive, you're going to change; if you don't, you're going to perish. It's as simple as that . . . you will not practice architecture, if you are not up to speed with this . . ."*

While this addressed BIM adoption, it is still most relevant today, because the industry is fragmented, conservative and most unproductive. But it is not all doom and gloom, many innovations and developments have heralded new methods, scope and made strident steps to make buildings better, leaner and operational performers. The procurement of a building should not end with its handover, but rather should open an embracing relationship for its entire life and beyond.

Such an arrangement requires marshalling, highlights skill-gaps that need to be filled and raises the bar with thresholds to be reached and completed, without missing a heartbeat. The competences involved delivering this seismic shift might appear daunting, but help is at hand. An overbearing word is blockchain and often is held at arm's length, but with closer examination much of the hype can be massaged into meaningful material.

Don Tapscott likens the situation to that of the internet's evolvement, which brought us 'e-mails, the world-wide web, dot-coms, social media, the mobile web, big data, cloud computing and the early days of the Internet of Things'. Essentially, this was an internet of information.

What is needed now is an internet of value, with assets becoming interactive. This includes identity and reputation, money, intellectual property, contracts, assets and energy. So, the matrix is moving from being squared to being cubed, adding a totally new dimension.

Malachy Mathews sees the combination of BIM and blockchain as evidence of value transactions. He claims this platform will disrupt the design and construction industry, and that data, as a commodity, has value. This is the crux of the matter, and that BIM-based collaborative technology will see merely self-serving contracts consigned to failure. This is to be endorsed and encouraged, and it will see hierarchical structures replaced by networked ones, becoming more efficient, enjoying higher valuations, being fault tolerant and self-regulating, through machine learning.

Construction professionals have often strived to recover the intrinsic value of their labour. Blockchain offers a new value proposition to extract reward not just for the collaborative services but also for the intrinsic intangible value across the life cycle of a facility. It becomes the gift that keeps giving, a contract that rewards value, a contract that does not reward non-performance (which is just as important) and an agreement that releases payment when due or expected.

The intension of this book is to demystify the layers and levels being amassed upon us today. The aim is to clarify how we have gotten here and what needs to be done to regulate and adopt these new paradigms. It has been my great pleasure to write this book and I hope you enjoy reading it. It is intended to bring a professional handbook aspect to the subject and a referenced content manual to dip into when needed. Each chapter will attempt to be concise and complete within itself, so that it can address problems and issues to respected areas of interest. While the entire book brings an overall treatment of the topic, blockchain is changing at a rate of knots which will be tracked throughout the timeline. I hope you like it.

List of Figures

Biography

James Harty is a lecturer at KEA, Copenhagen School of Design & Technology. His PhD researched "*The Impact of Digitalisation on the Management Role of Architectural Technology*," (Harty 2012). James was instrumental in the school's adoption of BIM in 2006, with the implementation of collaborative methodologies leading to deal with disruptive technologies.

He works with Common Data Environments, mapping virtual worlds on to reality and making digital twins relevant to Smart-Cities and Off-Grid, using Blockchain, IoT and AI. He has co-authored a book, "*Getting to Grips with BIM*" (Harty et al. 2016), and he sees BIM as instrumental in tackling climate change, sustainability, especially embodied carbon, and performance-based design.

His Master's in Urbanism mapped the housing types and urban morphology of a satellite town near Dublin (Dun Laoghaire), and he was a co-author of a report preserving "*Temple Bar*" in medieval Dublin. He is currently researching a Horizon 2020 project, "*ARISE*," implementing Blockchain and construction in blended learning across Europe, addressing all sectors in construction.

CHAPTER 1

Executive Summary

What is disruptive, fragmented and works with insanely minimal profit margins? As I said at the BIM Coordinators Summit, even the dogs in the street know the answer is construction, an industry which should know and do better. Even in terms of productivity, the sector is performing worse than back in 1960, because players can and often do get paid twice for double work, knowing that work will be demounted, in order that underlying work can be completed, before the work is assembled again. Making below cost bids also leads to needing requests for information and delays to lengthen engagement and encourage litigation.

What is needed is a method to actively engage stakeholders, even beyond handover, and more importantly to reward such an endeavour. If there is an incentive towards continued engagement, the benefits and potential is breath-taking. Key to this is performance, and key to performance is measurement. If a project is proposed that will save

Transforming the Construction Industry with Blockchain: Enhancing Efficiency, Transparency, and Collaboration, First Edition. James Harty.

20% in energy over the next 20 years, a method is needed to avert green-washing, to deliver the goods.

If such a situation existed, it creates a method to reward this performance, by say, paying a 5% of that saving to the perpetrator, as a reward, for making the building that was needed both for the client/user and society in general. A repeating paid-out dividend is an incentive not known in the industry today. Once established better practices prevail, making performance a central pillar in the mix, improving the sector. In order to verify and validate such endeavours, blockchain enters the fray. Blockchain offers a trusted framework for data validation. It records performance in a decentralised, immutable open-sourced manner.

To implement this new environment, smart contracts are needed. They use *if/then* structures to administer the work. They provide protocols that verify, simplify and enforce performances. Once met, they trigger a payout to the recipient, rewarding their successes. They also promote such activities and identify best practices. They change how buildings are made and they bring life-cycle assessment into procurement.

Automating this process opens the industry to integrating the Internet of Things (IoT) and artificial intelligence (AI). Digitalisation also brings common data environments (CDEs) and cloud computing, together with virtual reality (VR), artificial reality (AR) and mixed realities (MRs) into the new paradigm. AI introduces machine learning (ML), which will have a profound effect on proceedings. It can learn from analysing data and metadata and draws ever improving explicit data mining, as it is applied to large language models optimising and completing tasks unsupervised, freeing up time for other activities.

While ML is a subset of AI, it also gives birth to deep learning (DL), making expert systems, that can begin to think for themselves. It can identify potential bottlenecks and foresee mismanagement or toxic procedures that can be addressed before they become hazardous. This will transform the industry and make it a profitable, healthy, positive sector. Keeping tabs on this new paradigm requires a single source of truth and here blockchain comes to the rescue, ensuring probity and controlling procedures so that poor procedures become less attractive, being eventually phased out.

Finally, as Paul Doherty said at BIM Coordinators Summit last year: 'AI will not take your job, people using it will'.

How Digitalisation Has Been Adapted into the Construction Sector

If paper gave us the frontal architecture of proportioned edifices placed in the landscape marking territory, digitalisation gave us an additional aspect of parametrics. This brings a need to analyse and rethink data flows. Covering the work phases, actors, flows, covering, tasks and addressing the problems of these processes, skills and technologies.

Arto Kiviniemi has also drawn a comparison that in nature material is expensive while shape or form is in comparison free, while in traditional construction form has been expensive while materials are cheap. It is therefore most likely in the future form will be cheap and materials expensive, meaning that (3D) printing complex forms will not be a problem (Kiviniemi 2015). To illustrate this, he points to the forms of fractal shells for complex forms, modern monolithic offices for cheap shapes and the likes of the Olympic Bird's Nest stadium in Beijing or the Disney Concert Hall in Los Angeles as parametrical wonders of the world.

In building a case for change, Standards Australia makes a strong case that the construction sector forms a major part of the national economy (Standards Australia 2023). It quotes the sector as accounting for over one million people or 10% of the total workforce, generating over $360 million in revenue at 9% of Australia's Gross Domestic Product (GDP). Finally, it reports that every million invested contributes $2.93 million in output or $1.3 million of GDP.

Handheld devices have become ubiquitous on building sites and their use is becoming more and more entrenched into many practices, whether it is confirming completions, requesting information, reading QR codes, and even providing learning environments to access instantly, solutions and methods so that there is a readily available solution to potential bottlenecks, while reducing incorrect actions. They record situations and can instantly fetch the latest data or information or identify mistakes and can allocate the remedial action, saving time, effort and waste (Figure 1.1).

How Data Will Drive Generative Architecture Quicker Than We Think

Each wave of a new technology is usually heralded as a time-saving feature, leaving more time for design. This new void typically gets filled up with the protocols needed to implement these new competences.

Fig. 1-1: Student project (Bryan Zou, Dominik Wawrzyniak, Guilia Perciasepe, Oliwia Mazurek & Wikoria Ekiert).

Fig. 1-2: Student project (Bryan Zou, Dominik Wawrzyniak, Guilia Perciasepe, Oliwia Mazurek & Wikoria Ekiert).

Automating this becomes a game changer. We are on the cusp of generative architecture where function becomes a mechanism to derive a design. Computational algorithms analyse and explore possibilities based on specific goals or constraints.

So instead of consulting a book of Architects' Data, such as Neufert, the programme does this donkey work. For example, designing a school for 400 students, the number of classrooms, washrooms, shared and ancillary spaces can be generated automatically, leaving the designer time to mass and form the project to other parameters (Figure 1.2). And this is only scratching the surface, robotics and 3D printing could automate much more of the process, computer numerical control (CNC) controls machines by means of a computer meeting specifications by following coded programmed instructions without manual intervention.

Programmes exist to parse building regulations, to control delivery and to check compliance throughout the procurement process and beyond. Applications and robotics, better known as APPs and BOTs today provide an array of handlers to aid and abet a whole range of things. What brings all these things together is blockchain, and it is what coordinates and holds all in perspective in this new paradigm.

As Don Tapscott claimed in his TED talk in June 2016:

> '*The technology likely to have the biggest impact on the next few decades has arrived. And it's not social media, it's not big data, it's not robotics, it's not even AI, and you will be surprised to learn that it is the underlying technology of digital currencies like Bitcoin. It's called the Blockchain*' (Tapscott 2016a).

CHAPTER 2

Data Handling

Undoubtedly, data collection is the foundation for making better decisions to drive improvements, but it becomes complicated if there is too much data and it is not in an ordered fashion. Questions about ownership and management also make it difficult to navigate and its filtration and usefulness come into play as we wade into the melee. It can also grow into a data lake where in future it can be used in predictive maintenance. Data handling also requires an agreed environment in which all stakeholders are considered and have an agreed protocol for the exchange and use of the data generated through a project. It is built over four phases in which:

- The first sets out the processes, where the scope and content are broadly outlined for each stakeholder. This means it is an internal process, setting one's house in order before engagement.
- Second is the sharing phase, where information content is considered complete for some disciplines (those closely related

Transforming the Construction Industry with Blockchain: Enhancing Efficiency, Transparency, and Collaboration, First Edition. James Harty.

to each other) and is subject to review and modification to dovetail with the other parties.

- Third is the publishing phase where the endeavours of the first two phases blossom, and the data is definitive but subject to revision. In this phase, it is operational.
- The final phase is archival, meaning it is kept for the record and is accessible or valid but is now superseded.

In general, it can be seen as a fluid process, but with defined framing and defined channels of communication. By definition, it is easily accessible to all stakeholders who must each have a position and responsibility towards the content. It provides a method of traceability, meaning accountability is embedded in it. It must cover all data formats, which have to be agreed and understood, and it must contain open protocols for the exchange and accessibility of data. It has to be updatable and be confidential and secure.

By having such a model, then the flows and previous bottlenecks are made easier. All stakeholders can easily engage and be content that the data is up-to-date and accurate. It replaces the requests for information (RFIs), which could seriously delay a project, due to the time and analogous nature of the information flows. Typically, in the past, a request for information might only happen at a fortnightly site meeting, meaning it was first addressed at the next meeting and approved at the following meeting. This process when it becomes digital removes much of the downtime and the waiting for approvals.

It removes much of the paperwork, often taken up an entire day at the end of the week, which was dreaded and derided, with sayings like *the job's not done, till the paperwork is done*. Now an item of work can be allocated, when completed can be shown as finished (completion photo or similar) and approved for payment all within a cloud solution. This digitalisation process is revolutionising the industry, and once adopted is rarely rescinded or withdrawn in favour of the older methods. It is controlled by an International Standard, ISO 19650 (previously being the PAS 1192 series) and manifests itself through many platforms. These include but by no means are exclusive: Autodesk's BIM 360, Dalux Field, Box Delivery and FM, Revizto, ASTA Progress Mobile App and Field View.

On the construction site, in order to ensure proper coordination and dissemination of tasks SISK, an Irish contractor, involved with Center Parcs leisure centre, used digital project delivery (DPD) methods divided into fourth dimension construction sequencing, building information modelling, a progress mobile app (ASTA) and another called field view to keep everything up to scratch. Field view was used in their quality sign-off procedures, to ensure environmental and safety auditing on a hand-held device, reducing paperwork and waiting time for approval. Not only does this improve the day-to-day management of the site but also it builds into a bigger picture where things can be monitored, best practices gleaned and improvements made on evidence-based decisions, making the whole enterprise better and fit for purpose.

This reduces time spent using paper formats, it allows for tasks to be easily raised in a meaningful way while saving time with inspections. Tracking live progress was critical for them on the project, using the ASTA mobile app, a planning software which allows GANNT charts to be prepared so that informed decisions can be made while reducing the amount of data inputs previously made manually. This improved and increased programme awareness throughout the entire team made data entry faster, increasing value added time for better planning.

BIM, especially 4D BIM was instrumental on site to utilise and plan the works effectively. Their software allows for the collation, distribution and management of all of the project and technical documentation. This means that their designers can off-feed their information through one platform allowing collaboration for the various stakeholders across the project. It is a key delivery tool that has been around a while but in combination with the other tools adds to the overall success to the project.

4D modelling has been used on site in conjunction with the handling, the logistics, the supervision and day-to-day delivery scheduling and the call-offs from their suppliers moving outside of the site and involving third parties in a lean process that could cause multiple delays previously. A reluctance to implement such strategies was quickly dismissed as the benefits accrued during the project as workers saw the improvements to workflows and reaped the paybacks in their workloads with improved efficiencies.

Maximising these activities on site allows lean constructions to improve work sampling, just-in-time (JIT) delivery and something they call 'kitting'. Kitting involves the correct quantities being sent out to the correct assemble and installation points. In their many on-site warehouses, it is used to utilise the kitting. Essentially, a kit of parts is put together for each and every job, just like flat-pack constructions that come with the correct number of parts and the correct number of fittings, needed to assemble it in-place.

This reduces wastages and promotes collaboration, as when it is bagged, boxed or palleted, the receiver knows that all that is needed is supplied and does not need to be quality checked and accounted for, improving cross platform collaboration for all concerned. So, it comes to the right location with the right quantity. For example, a boiler kit for a particular lodge. This means that the fitter receives the right quantity and the right specification in the bag, eliminating waste while improving the workflow.

If kitting is a method used for management of materials, then work-sampling is a method used for management of resources. Sampling is a method of observing activities ongoing so that they can be classified. Basically, a worker is assigned to monitor blocks of work lasting an hour where all activities are recorded as value added, a support activity or waste. Following the exercise areas of waste are high-lighted and reviewed to see if they can be eliminated if possible. This might include a method of detailing to make the task easier, to make some activities simpler or quicker and to look how the crews were set-up managing materials and accessories in the field.

The benefits of work sampling in one instance meant that they were able make a 22% productivity improvement observed in key trades, bringing packages back within budget. So, in a fast track, large-scale unique project with its own set of key challenges, the team has used cutting edge technology to help deliver this project in an efficient and effective manner. This means that they remain on-time, on-budget, delivering to the highest level of safety and quality. Embracing technology can be a challenge at times but their future success depends upon it, says Brian Kennedy of SISK contractors (Kennedy 2019).

So, the research question would be: What is the process that is being replaced? What are the advantages of implementing such a regime? What are the benefits of adoption? Where does CDE lead stakeholders to in the next phase, BIM Level 3, Building Lifecycle Management? What is being replaced is inefficient methods that support all the problems that have beseeched the industry. The advantages encourage better collaboration, remove double work and improve productivity. It fosters a better method for better practices, and it can reward such practices.

Having a single source of files, typically in a cloud or, in a common data environment (CDE) means that there are no revision mis-matches, that files are always up-to-date, and accountability is easier to handle as there is trail of who has done what and where, it is traceable. Also, as BIM Level 3 expands, it can generate a digital twin which means that the project is built virtually before it is realised. This means that errors and erratic solutions can be caught before stepping on site. The twin allows *what-if* scenarios to be confronted, and this enforces facilities management issues to be addressed and tested. The twin replaces the Facilities Managers (FM'ers) making a fresh clean sheet report after handover.

Data Mining

If data is ubiquitous (and it is), then feeding artificial intelligence (AI) more and more data becomes an endless thankless cycle. Moreover, the AI's appetite becomes voracious and grows out of control within the realms of human cognisance. 'The AI dilemma' (Harris and Raskin 2023) discusses that 50% of AI researchers believe there is a 10% or greater chance that humans will become extinct from our inability to control AI. Harris and Raskin outline that new technologies need to be responsible for their consequences. This is akin to engineers of an airplane telling you there is a 10% chance that the plane they have created and of which you are about to board has a 10% chance that it will fall from the sky with you looking out the starboard window (ibid).

Pointing to how social media got it wrong with fake news and similar and the need to be allowed to be forgotten, and the need to protect data, which were all unforeseen consequences of the new social media rolled out before their full impact was known, they now point to ChatGPT and how it will be unleashed on the poor unsuspecting public again. But now they point to an exponential rise and even a double exponential rise in how AI is using machine learning to create a monster.

The ills of social media are numbered as: information overload, addiction, doom-scrolling, influencer-culture, sexualisation of kids, QAnon, shortened attention spans, polarisation, bots, Deepfakes, Cult factories, fake news and the breakdown of democracy. AI, they see as compounding the following: reality collapse, fake everything, trust collapse, the collapse of law and contracts, automated fake religions, exponential blackmail, automated cyberweapons, automated lobbying, biology automation, exponential scams, A–Z testing of everything, synthetic relationships and AlphaPersuade.

This is off-topic but unnerving none-the-less. Some of these are self-explanatory but some are totally unheard of. The last, AlphaPersuade, draws my attention, in that it will relentlessly bombard me in all manner of means, to sway me over to a totally different set of values. This is in effect Deepfakes, that will relentlessly pursue my values and impose their views above mine, until I either accept their content or in despair give up my own values in dereliction of my own disposition.

They predict the end of elections because of the polarisation of voters and the beginning of this could be said to have occurred in the recent US presidential elections.

ChatGPT is a chatbot using generative pre-trained transformers (GPTs) to accumulate data from many sources. It creates a language. The transformers are text, images, etc. used to predict or create a new one. This is called a generative large language multi-model but where, they maintain, content-based verification ceases to be created. A machine they claim does not have a sense of framing or morals to know whether something is good or bad. They cite the instance of a 13-year-old girl asking her personal AI on Snapchat about someone grooming her for sex with absolutely no warning signs of the inherent danger being raised by My AI, in fact scented candles might make it a memorable first night, she is told (sic).

They call for rules of engagement, primarily that new technologies require a new class of self-imposed responsibility; that the new powers released in these technologies create a race to maximise engagement in the engagement economy (wanting to keep a person on the app for as long as possible), and that without any coordination that such an endeavour is doomed to tragedy (ibid). They see AIs as having a bias without transparency.

Conversely, Joe Lubin, co-founder of Ethereum, sees 'some kind of a truly decentralised autonomous organisation owned and controlled by its non-human value creators, governed through smart contracts' as going 'All the way' towards decentralising the enterprise with blockchain providing a frictionless efficient mechanism to run the show (Tapscott 2016b).

Vitalik Buterin, who created Ethereum, imagines a bot that could roam the internet with its own wallet, learning, adapting and developing a taxonomy penultimately leading to full AI (ibid). So, Don Tapscott sees blockchain as being the keeper of law and order, and as having the transparency so that any such shenanigans will be exposed for all to see (ibid). Admittedly there are a number of years between both positions and one which is changing at a rate of knots, to which I shall return.

Data Communication

Building information modelling brought with it a whole host of procedures, processes and methodologies about how we interpret the data generated. What is this data and how does it communicate across the platform are two interesting questions to be raised. Data conveys information about quantity, quality, status, statistical with meaning or sequential impacts having value.

It communicates with each other by having a tagging system, identifying each piece of data with a classification system to be managed across the platform. Ownership or authorship are also two very important factors in generating and valuing input to the project. From this remuneration and rights can be assigned to all stakeholders and third parties.

A typical classification system will identify the elements, usually with regard to where they are in the project. A coding process takes place to give a unique tag to each building part or process. This usually involves the tracking of metadata associated to each element to document it in the project. This data includes (but not exclusively) project id, organisation id, project title, business type, knowledge, discipline, contract, procurement, work area, work phase, file type, storey, sheet name, level, information, purpose, exchange and status. Under knowledge differing types are defined including architecture, urban and landscape design, interiors, industrial design, surveying, information and communication technology, mechanical and structural engineering. So, in a Danish system (bips, now Molio), the following code would have the following definition:

|4_A6_K01_H1_E1_N01

> *Project ID ('four' from a known list), 'Main Project' Workstage, 'Architectural' Discipline, 'Plan Drawing', 'First Floor', Running Sheet 'Serial Number'. This would be entered on the drawing sheet title block and be unique within the documentation.*

Other classifications include Uniclass (UK), Omniclass (USA), SfB (S) and DBK (DK). When federated models from differing stakeholders are set together, methods are needed to translate and interact with each other while accepting each of the other systems. Most of this happens on a like with like basis. Problems only arise when a field is not catered for and so translating back returns a void. Industry Foundation Class (IFC) is heralded as an open-source format to which all authoring programmes can write to but from which no authoring can occur.

Moving a simple wall in and out of several programmes can lead to data being dropped. Typically, a field would have no corresponding field in the new format and if not critical would be dropped. On passing back that field would be voided (Pazlar and Turk 2008). Even using IFCs, evidence has been shown that all export functions were not supported. Pazlar found that something as simple as a wall hatching or pattern being lost in a vertical section. This puts the onus on the operator to be vigilant, not blindly trusting the mapping process (ibid).

IFCs, while offering a common base for all stakeholders are often disliked by vendors, as it allows competitive software platforms to match and offer compatible solutions to each other, and in so doing offer alternative viable results. So often, with a new revision of a software platform, a new feature is not matched or covered in the pooled information, giving them a temporary advantage being unsynchronised until rectified. This time difference is exploited to the major vendor's advantage, and countered with this is advancement and not doing it would be regressive.

This hustle and bustle, both from vendors all the way to users, causes everyone to be on their toes, and this leads to Information and Communication Technology (ICT) protocols where all is agreed and is binding at the outset of a project. They are unique to each project because of the makeup of each and every project team. They are also dependent of the client, being a smart player or dumb, wanting the capital cost to be kept to a minimum, without understanding the intricacies of this new paradigm. Because they are agreed at the outset, they often miss nuances in the project and can become redundant in long running projects.

The complexity of this new layer has meant that when embraced the early adopters are happy to sit on their laurels and become somewhat hesitant towards continued expansion, in the belief that they are holding the holy grail. This means that BIM Level III is not being adopted and rolled out as was expected and anticipated. All in all, this can be seen as disingenuous at best and limiting in best practices at worst.

To correct this situation, certain automations need to come into play. First and foremost a coding to translate and match each and every element in the project. Indeed, when opening a project, the classification system could be chosen, independent to the other stakeholders. These choices can be recorded and noted for subsequent project and a taxonomy can grow and be beneficial to similar projects. This can also apply to stakeholders where teams that work together in similar operations can predetermine many features so that an intimacy and familiarness grows with like-minded participants. Conversely, a new contributor can rack up a project to their requirements so that they can compare like with like. All circumstances are covered.

Secondly, these protocols become invisible, removed from the initial pitches so that energy can be focused on other aspects of the job. This frees up much effort and stress for other challenges of the project (BIM Level III, for example). Control of these additions can be tackled in the common data environment (CDE) so that whatever handheld device used can be tailored to the users' preferences without detriment to the rest of the team. Delegating so much authority requires a robust platform, which needs to be verified and vigorous. The app running all this is a form of Blockchain and will work in the background, seamlessly with all the other players.

Data Filtration

Because data is so pervasive, controls are needed to make it compliant with our wishes and how we want to use and engage with it. This requires a form of filtration to bring relevant facts to the fore and prioritise the worthiest of them for the case being examined. Essentially, what is needed-to-know and nice-to-know. This is not to say or affect the cohort as is, but to bring a sense of order to the proceedings. How this is managed and who decides on the issues become most compelling. There needs to be a hierarchy and a sense of relevance so that the convoy is not drawn to the bottom of the seabed, drowning in a data overflow.

Many procedures and data streams have been suggested and implemented. But this only opens a plethora of differing protocols which might or might not be transferable or translated across the domain. We are back at the classification system again, but with a new angle, relevance and probity. There is now a need to be knowing of what the subject or object requires and to be all-knowing in delivering this quantitative master class.

Procedures and methods have been made and handed down through the ages by those who have worked and generated this data, but now we are entering an overhaul of data's relevance and metadata's impact on the areas of interest and scope of the projects in which we are involved. There is a danger of flooding the domain and of making the same mistakes twice, not learning from past encounters. This latter point is most valid when it comes to repeat work and the benefit of hindsight in learning new situations while being in command.

Experience has great value, but in an ever-changing world in which we live, it can be too expensive to pay for, and so other methods are deemed necessary to complete the gaps in our skill sets and how such a set is acquired. Shortcuts sound as if corners are being cut but if it results in qualified decisions being made with robust results, then what is not to like.

Fast forward and automate these issues and a formidable enterprise emerges. Fundamentally, it identifies the most relevant pieces of data, removing redundant stuff so that qualified decisions can be made. Furthermore, it might also implement these things so that the user is freed from a paper trail that could be fatal in its workload and demand. In such a scenario, the process proceeds trustfully to the stakeholders with a controlling sense of euphoria not seen before.

This whole effort will and can be augmented by blockchain. We do not question google maps when they show us the landscape. We accept many apps per se, without thinking about the consequences. If it then ranks and rates the various modes of transport, if we seek directions, it will inform us the best method. This is blockchain at a basic level. We do not flinch at this information and accept it untarnished.

Digital Control Room

A digital control room could be described as a war room, in that it is a command room, which is the epicentre where co-ordination and collaboration takes place while a project is in procurement. It can be used in design phases as well as construction itself. It is a place where all issues and hot topics can be highlighted to relevant stakeholders. It balances out any discrepancies, where progress can be broken down to each trade, in order to manage a project efficiently. The workflows allow a lean operation where issues and root causes can be brought to the attention of the design team very quickly. It brings many JIT features to the fore.

It also brings together an enormous visual aspect so that you can manage and resolve all areas of the project. It comprises of a digital whiteboard which has a physical size of up to 4 m, with a virtual length of up to 90 m. Paddling motions of the hand push and pull the pages back and forth. It is populated with Gantt charts, excel sheets, documents and models, which can be interrogated in real time. Integrated Concurrent Engineering (ICE) plays a big role here. ICE is a social method, helped by technology, to create and evaluate multi-discipline, multi-stakeholder virtual design and construction (VDC) models extremely rapidly in real time.

Where it will transform itself is with blockchain. Suddenly a list of to-dos can be vanquished as automated apps and bots complete tasks. Again, I refer to low-tech, as we might call it today, including all Internet 3.0 situations. These include Internet searches, Google rankings and Facebook algorithms, which furnished a whole generation with facts, and pumped them with add-on features to extend their attention spans and promote extended engagement. The bottom line invariably was a cash incentive that primarily matched sellers with possible buyers. The incremental possibilities went unnoticed with the general public, but it made multi-billionaires of the owners of such platforms. This was totally unprecipitated by both governments and the populus in general, and subsequently much damage was done.

Data Is a Commodity

Data is a commodity; therefore, it has value. Crass as it may seem, it is most compelling. Once accepted, it can become obsessive, in that the possibilities and potential explode into reinventing new business models, rebuilding and changing the boundaries in which we operate. It creates an immersive place to develop the project, it provides a place to showcase aspects of the project and it allows all stakeholders to explain how they got to where they are, from a striking standpoint.

Data has also been called the new oil. It powers many of today's transformative technologies, feeding AI and automation, while predictive analytics is growing enormously. Data in its raw form is largely useless, so it needs to be cleaned, sorted and refined to reveal the information needed for strategic business decisions. This begins to draw attention to its quality, and appropriateness.

Design in the built environment generates a lot of data, which needs to be appreciated and analysed properly. The data associated with this process informs specific design solutions driven by multiple, usually competing objectives that need to be taken into consideration during fast review cycles. The fast review cycle is a new phenomenon not previously encountered live on the fly, but rather racked up through hard experience and always in hindsight.

Data can be as simple as floor areas, to more elaborate metrics such as thermal performance, carbon footprint or contextual integration, derived by a plethora of time-consuming analyses (Kosicki et al. 2021). Previously, a building's agenda was merely functional and aesthetic, but increasingly energy, CO_2 and sustainability are pushing performance to the fore with qualified metrics. So how can we deal with this, in production, classification and hence filtration and finally the reuse of these findings for future projects.

Through handheld devices we, the general public, have generally surrendered or waved our rights to privacy of our data and so in exchange for *free access* to platforms, data has become a currency today, driving advertisements and empowering third parties to bombard us with unsolicited feeds. AI stands as major player with big potential to embrace architects' knowledge and skills, enabling faster design analysis and expanding creative capabilities.

The first approach involves the application of surrogate modelling to replace conventional and time-consuming modelling processes with economically efficient predictive models. These models are spanning from intelligently simulated results and quickly detected design within historical data, therefore enabling quick and informed decision-making. On the other hand, second approach introduces design-assisted modelling, ideal for being integrated into designers' thoughts and ideas to facilitate architectural processes that are missing analytical cycles (Rajković 2023).

In architecture, datasets exist in different and various file formats. AI, on the other hand, is equipped with image, graph, text and voice-based tools capable of searching into digitally archived data from thousands of projects and file formats. AI efficiently identifies data relevant to specific design task, leading to significant time and cost savings (ibid).

Key features of AI to the construction domain include optimised construction planning, contractor and subcontractor selection from tendering procedures, smart construction site management, health and safety management, advanced technologies implementation, waste management and sustainability. These contributions in the construction sector represent a significant shift towards enabling efficiency, safety and sustainability within the industry. They reflect AI's potential to revolutionise the construction sector and help its environmental impact while optimising project planning and execution (ibid).

AI is invaluable to address challenges in complex or labour-intensive tasks. This covers automation, robotics, risk aversion, efficiencies and last but not least health and safety. AI can extract valuable learning from the digital process. Its position in the critical mass means that it can act swiftly to recover just-around the-corner failures and analysis of the data both now and historically to plan and plot better engagement across the whole enterprise. It can deliver more intelligent and informative reflections to generate new reasoning and tacks. This will lead to improved project efficiency, better quality, rewrite collaboration and meaningful sustainability.

Project management is the 4D and 5D of modelling, namely time and resources, together with quality, to meet the wishes and requirements. But there are soft issues around each and every project, such as the pervading economy, the reigning political regime, physical and social factors which can influence any and all decisions in a project. Each project can be altered by outside concerns such as supply-chain delays, workforce recruitment and poor reworked effort.

So, managing the available resources is paramount, addressing conflicts in scope, cost, time and quality are always to the fore. Discovering these before they become serious is in everyone's interests. AIs ability to match human behaviour with learning algorithms enhances a new platform beneficial to both parties. Whether AI makes the decisions or presents them to the stakeholders is currently in its infancy, but just an automatic self-driving vehicles are getting better at negotiating the road network due to the magnitude and numbers of vehicles collecting the data leads to better cognitive decisions, so too will its impact in construction.

This will also affect code compliance, as rules and algorithms learn best practices, so that they can be patched into current practices. Collaboration will also be aided as AI can package deliverables into better forms of exchanged data. By this is meant that formats and classifications can be aligned automatically and seamlessly to all stakeholders, by optimising the process, enabling the optimising of the data into formats expected in each disciplinary silo, which currently is dealt with using ICT contracts offering the least form of resistance.

Similarly, robotics will bring new scenarios to the building site reducing the workers to exposure to dangerous environments and activities. This will also be driven by the shortage of qualified professionals in the digital world. The EU us undertaking extensive programmes and platforms to address this problem, but it means having portals and applications for training and retraining both the blue- and white-collar workforce. But as Paul Doherty said at the BIM Coordinators Summit in 2023:

> *'AI will not take your jobs, but people using AI will'*.

This was in his presentation: 'Unlocking The Metaverse: Digital Real Estate, BIM and the AEC Industry' (Doherty 2023).

In the adversarial economy (adjudicated by the legal sector), AEC professionals have been incentivised to minimise the transfer of information between parties. This is counterproductive. Blockchain has the ability to affect even smart contracts, that is to say, that if there is something to be done when it is complete, it can be appraised in real time, and payment can be made and verified.

Blockchain uses cryptography to create a trusted framework of data. This is called a ledger. Blockchain also means that the records are decentralised, immutable and readily available. Blockchain can be used to:

- Establish and verify identities,
- Record transactions,
- Register and track assets,
- Share information and more.

An important benefit of blockchain is that it creates a single version of the truth, thereby eliminating redundancies, outdated records and conflicts. It also allows organisations to improve trust, efficiency and the user experience without replacing legacy systems or losing existing data. Most importantly, it can validate. This is a backend feature, which allows employees and employers to remove a painful part of the hiring process where letters and paper versions of documents must be supplied and verified.

Blockchain offers three elements.

- First, blockchain has a trace and traceability, a real-time method of showing where the student is on their learning path.
- Second, it can be a ledger, noting what a person has learned without being compromised.
- Last, it can reward such practices with a coin that the student earns for completing modules, guaranteeing evidence to would-be third parties. While sounding relatively nominal and simple, this ability is intrinsic in a method needing transparency regarding demonstrating incorruptibility and robustness that stands up to scrutiny.

The competitive book named below, *'BIM and Construction Management'*, combines theory and practice in equal amounts to both direct the reader and stimulate him/her as to why. This engagement is a good balance, becoming the go to book on the subject.

I was a co-author of 'Getting to Grips with BIM' which covers the theory, the practice while being relevant to small medium entrepreneurs (SMEs) who most needed it. I would hope to bring these features in a new book.

How Information Can Be Filtered and How Blockchain Will Feature Herein

With regard to contracts, there are three basic essentials to their creation: agreement, intention and consideration. So, both parties must agree, there must be something to be done and there must be remuneration for the said work. Automating this process, with a view towards smart contracting, would see a movement away from analogue content towards digital means. This means that tender material would move from drawings and descriptions to modelling. So previously, the work would be registered and written down, now it is embedded in the model.

The new scenario would see all the work coded and priced in the model using blockchain. By definition, the model must be so detailed that it constitutes a *de-facto* digital twin. The building elements, or objects, which are not in the model, such as frames and panels, screws and nails, are usually covered for calculation work by rules and specifications. So, the options are to lose these contingencies or to model all elements in the project, which could be very expensive. In the bigger picture take-offs would need to be covered in the price books and attention would be needed to address this new consequence.

Furthermore, if additional work should arise (that is, not modelled) or only found when discovered in project review, the client creates a new agreement note (a new smart contract) for the requested work, the contractor would then accept it so that the work can be coded into the model through the blockchain and subsequently carried out, the payment can be executed as a smart contract. All deadlines are removed from the contract, and it only works through the model. This removes unwanted payments through the smart contract and the problem of the contract expiring and the automatic write-down in the guarantee taking effect (Laustsen 2021).

Information Revolution versus Intelligence Revolution

We all know that AI is neither *artificial* nor *intelligent*, rather that it is *real* and *dumb*. While this is an exaggeration, there is some truth in it. It is certainly a reality, and it is here to stay, whether it is a help or hindrance is up to discussion. The fact that it can generate deep fakes is worrying, but the fact that it can generate agency is enormous. By agency, I mean, the ability to opens up new veins of thought and innovation, while progressing human endeavour, which can never be underestimated. It brings new eras of trust, new opportunities for research and begins to leverage the platform for new applications.

Edge computing is a strategy where smart devices collecting data manage how and where the data is handled. The edge focuses on how the data can be gathered and processed on site. This reduces the amount of time sending and receiving data back to a datacentre or the cloud. It means that it can be analysed in real time too. It creates a faster more reliable service, supporting latent applications while identifying trends and patterns. The improved response creates a user-centric and energy-efficient building with their associated cost savings.

Proactively, Delotte and their Dutch headquarters, The Edge, in Amsterdam, is a case study showing the implementation of occupancy and dashboard monitoring of the facility, through their 'Power-Over-Internet' lighting which acts as a two-way street collecting data and reporting the facilities demands, making it the most sustainable building in the world (van Oostrom 2016). Employees basically revoked their right to privacy by letting the building manager have access to their location in the building. This means they can be tracked at a meeting, in a quite working space or using the toilet.

Digital Impact

Metadata is data about data or data that describes other data. It can be structured or unstructured. The prefix meta typically means an underlying definition or description within technological circles. It can be latent, meaning that it happens unknowingly. Typically, it is known to those who need it and so, by definition needs careful filtration to bear fruit. This means that its management is vital to use it purposely. Therefore, its impact is essentially a new paradigm, charting new territories and opening up immense possibilities.

The template that this craves is more often than not through algorithms. These are sets of rules which calculate numbers or solve problems, especially using computers. It involves data processing, automated reasoning and similar tasks. In its simplest form, a quadratic equation would be an algorithm. This kind of comparison makes it tangible for my generation.

As an intern while at college, I received a project which had just been refused planning permission, because it overlooked the neighbouring properties. The project was four two-storey apartments in a single storey area. The problem to solve was to see if the same floor areas and number of apartments could be achieved in a single storey solution. The original proposal wished to place four long narrow fronted apartments into the confined space across the smaller dimension of the site, using a double height atrium to allow light into the deeper spaces. But it was two-storey, and so was refused.

The solution would mean using the longer proportion of the site and making the properties shorter and wider. There was also a requirement to provide a minimum garden. This suggested an 'L-shaped' building with the garden completing the footprint. The longer dimension divided by four did not use the site most efficiently as there was leftover space in the proposal. Having three projects beside each other and turning the fourth at right angles made better use but the relationship of length and breadth now was open ended and would require testing and testing to fine-tune the results.

So, there was a square meter value, an 'a' dimension and a 'b' dimension that needed to fit into the length of the site: A quadratic equation has the form:

$ax^2 + bx + c = 0$ *(where 'a' is not equal to zero)*

The equation resulted in the single storey house fitting three in a row with a fourth at right angles closing the space. This appeared to bring closure to the problem, and I was about to report to my boss just so, but looking at the site, I saw that a fifth house could be added in the excessive carpark, and so I placed an extra house on to the site. Bringing this to the boss's office brought cries of 'show me' followed with the lifting of the phone to tell the client he could get an extra 25% on the site.

CHAPTER

3

Trust/ Opportunities

If strategic alignments need levels of trust, what is trust? John Egan (Egan 1998) failed to define or formalise what is meant by trust or how it is recognised. Initially in open collaboration, there was a call for open communication. But Hedley Smyth reiterates that trust is not about open communication, saying that if there was complete transparency of communication, then there would be no need for increased levels of trust (Smyth and Pryke 2008).

Rather he defines trust as being a dynamic set of concepts that are present in its formation and development, with a philosophy underpinning this set relating to both the moral and economic issues, together with a methodology showing how trust is formed including its dependency and its management. This is then broadened into the characteristics of trust, the components of trust, the conditions of trust and the all-important levels of trust.

Strategic alignments objectively look at the revolution from draughting to modelling, together with the collateral that is necessary to implement such a shift. Draughting has always been a formalised

Transforming the Construction Industry with Blockchain: Enhancing Efficiency, Transparency, and Collaboration, First Edition. James Harty.

procedure allowing a creative person a method to capture thoughts and to convey them to another for implementation. It blossomed in the Renaissance bringing abstraction in the form of proportion into buildings. In modernism of the last century, it brought a new life to form in its absolute meaning. It is constantly balancing these two concepts of form and formality.

With the complexity of modern life, this process has evolved into a collective operation. The size of buildings invariably means that it is a team that prepares the documentation for the building and another that executes these instructions on site. It is this process that attracts my attention here, it has nurtured a whole industry that acts as a barometer to the national economy.

Where this is changing is the teams that are forming and growing to blur the polarities into a continuum. By pooling resources into a single model or a federation of models, IMTs allow management of these enterprises, administer risk and eliminate double work while diminishing human error. IMTs have changed the focus away from the technologies and directed attention towards the process, which is admirable.

A lack of designers and constructors working together is changing. Clark of Hilson Moran, a leading multi-disciplinary engineering consultancy, are doing a lot more work with contractors, where they are on-board early, and this is very much a two-way street (Waterhouse et al. 2011).

> *'We are designers, we like to think we know about buildability and construction, but the input we can get from contractors is invaluable, and that is improving our design work'.*

There is a cultural change:

> *'It is a different way of working. Gone are the days of where you have an engineer doing (only) engineering, where he does a mark-up and it goes to a CAD operator to draught it with a certain amount of engineering intellect . . . that's now different. Even in our office, you walk into an area that used to be the CAD department, now you have got an engineer and a CAD operator sitting next to each other, almost all of the time'.*

The major obstacle to this is, how trust is nurtured, how new blood can enter the mix and finally how information, competences and knowledge is shared for the benefit of the team, the project and society at large. Initially, there were calls for sharing or giving away data for free, but with contractual obligations and recovery of costs, there is a great reluctance to do so by the players and those who have invested so much into the project, to see others in the supply chain capitalise handsomely at their endeavours (Williams 2009). We have also seen in the case studies that contractors do not see ownership as an issue but that it is the model that is important not unlike the baton in a relay race.

How this can be remedied rests with the client and the appointment of all the stakeholders in the project as well as contractual concerns. First, principles say that work effort must be remunerated, and second, there cannot be subsequent adversarial disputes about the quality and correctness of the data. A designer cannot and should not be shackled into using binding contractual methods of procurement and application during the early concept phases, just as there has to be development in the process by the time production information is at hand. There has to be a *de facto* acceptance of the state of the data at each stage of the procurement and a method of improving or altering data if and when necessary, and there is, it is called a contract (Miller et al. 2008), but it is a new type of contract, citing trust.

Practically speaking, this can occur in two places; either within the model using the model phases property or in a viewer programme, which holds the diverse entities allowing them to be overlapped, collision tested or time-line compared. The 'within-the-model' option allows objects to co-exist in time and space without displacing each other, while allowing the data to be shared. This is a paradigm shift.

Within the free viewers, such as Navis Freedom or Tekla Viewer, there is an opportunity for many formats to be assimilated into the same federated virtual time and space, allowing many operations to be completed and reported. Integrity and ownership is not challenged but everyone from planner to environmental activist can access the data for whatever reason. Filters and views prepare the data in optimal sets for the users. This means that neighbours can check building heights and overlooking from their own homes that sub-contractors can check if there is room to transport their plant up into the cramped roof space or that project managers can check that in week 37 that the project is up to speed, that a bottle neck is looming or that, God forbid, that they are ahead of schedule.

The benefits of the model are not lost on some flexible entrepreneurs, already there are stakeholders who are entering into mutual agreements to work together and reap the rewards of completion on time and to budget. The biggest issue here is risk and how much or how well you trust your partner. Building trust in a business environment and especially in a fragmented construction market requires new skills and new procedures. Changing work practices from the adversarial to the collaborative requires major changes in mindsets and even social behaviours. There is no 'I' in TEAM, but there is a mangled 'me' lost somewhere within!

Leverage Devices

The cultural change required to implement integrated practice delivery is an enormous challenge defining 'true partners and collaborators with a mutual interest in a successful outcome' (Smith and Tardif 2009). Essentially, it alters the way and amount of time consumed in being adversarial and in expecting litigation. Increasingly, contracts are explicitly saying that stakeholders will not sue each other that future legal action is a 'no-value' task and that trust with verification mechanisms will become standard, as in banking. The principle cause of a bank failure is often a loss of trust rather than insolvency, there is very little difference between a failed bank and a health one, Smith tells us.

How this impacts technology is principally in the transfer of information and the risk it imposes on the authoring party who could be held responsible for the quality, completeness and accuracy of the handed over data. If a 'no-fault' policy is in place, each stakeholder accepts the data as 'found' and must validate it, appropriately to their needs. Validation consists of two parts, determining if the data is from a trusted source and confirming the integrity of the information itself. Smith calls this stewardship. Where there are errors or omissions methods will have to be effected to compensate the corrector or rectifier instead of identifying the responsible party or assigning blame. The blame-culture stagnates the process and causes delays. There has to be a hand-off of responsible control.

This greater dependence of stakeholders on each other can cause strain within the working relationship if trust is not present and more importantly earned. In order to minimise and in an endeavour to make the process, more transparent standards are invariably required. The National Building Information Modeling Standard (NBIMS) of America has deployed a compendium of principles called a capability maturity model (CMM) to define goals and offer methods of measuring business relationships, enterprise workflow, project delivery methods, staff skill and training and the design process against an index (Smith and Tardif 2009).

This allows for a form of benchmarking and acts as a quality management control for all those involved. It covers the data richness, life-cycle views, roles or disciplines, business process, change management, delivery method, timeliness response, graphical information spatial compatibility, information accuracy and interoperability support. But it is only a skeleton which can offer the stakeholders an index to measure or check each other out and to bolster their own pitch by giving them the tools to build their own argument and set out their own stall.

Increasingly, this tracking of data will become automated with machine learning hoovering up all these loose ends and laying them out in neat, ordered schedules. A hybrid solution will allow human intervention, but ultimately, they will be so trusted that there will be no human interfering required. This not to paint the human as an error risk but rather to give a platform that can be trusted and used more directly with faster results and safer procedures. It will also identify issues before they become costly or detrimental to the procurement.

Finally, it will remove the getting up to speed requirement of learning new procedures and protocols, which today makes a certain cohort indispensable. BIM managers' roles will change and the dependency of them to the smooth-running construction will become reduced. Digitalisation is here to stay and that will need tending but it will be more robust and resilient.

Light Touch for Agility

Essential to collaboration is the first line of the contract that signatories will not sue each other. While sounding innocent this is a major step. Methods have to be found to remunerate work at a fair rate. Competences need to be appraised and rewarded appropriately. Changes and error rectification need to be awarded to who is best placed to do the work. There has to be an incentive to complete on time and to budget. There has to be mutual respect for all in the supply chain, and this is called plainly and simply, trust.

This in turn is seeing the phenomenon of capability maturity matrices (CMMs) appearing, where differing parties tabulate their competences, their bond values and their ambitions or experience, and others compare and contrast it with their own so that strategic alliances can be formed. This is not unlike speed dating, and the metaphor does not end there. These collaborations are not for a single project but are related to the longer term. If a team comes together and competes and completes on a hospital (say), then they will try and corner that market and capture all related work.

Comparisons can be seen in large legal firms for architects, and also in major contractor/developer firms and large consulting engineers who feel they have the momentum to carry this off. But there is room for small players too and smaller targets but this is ongoing. When it filters all the way down to sub-contractors who can take off quantities from the model, then significant progress has been made.

Typically, these consortia comprise a design team (architect, structural and service engineer) who use or plug into the same model. From this, a surveyor or estimator can extract quantities from the model and together with a price book or work rates and material costs can price the work. Moreover, once each component is type coded it has a classification, which can be linked to specifications clauses to generate full building part specifications. Following this, a contractor or project manager can begin sequencing the work so that there is control on site with proper manning and resources.

During design, changing a material out for another is typical. It might be changing a wall construction from brick, cavity, insulation and load bearing concrete to a brick slip external cladding, allowing more room for more insulation. This changes the construction fundamentally and there will be knock-on changes to work's specifications, site assembly, detailing and cost. Within the model's properties such a change would automatically be switched and the change would be seamlessly altered.

This is a major revision to procurement. A student of mine started this process with the properties of elements in the model. When concrete was in the mix, all the concrete features were included. Changing to CLT saw a raft of changes reflecting the new situation (Palsbo and Harty 2013). In the demo as a wall was marked on a plan, the fly showed in real time the cost, the specification and all related data. Selecting the wall and changing a parameter responded in real time too. It meant that a traffic-light control for cost, carbon or energy could be integrated into the interface so that real-time decisions could be made to influence design.

A Sentient Machine

Parallel to this, financial viability is finding its way into building information modelling where computed area schedules are being mapped in early versions of the model which can be maintained and updated through the procurement of the project. Linking this to indexed price books ensures better cost control and improves project certainty. Facilities managers are also finding ways to map their requirements into the model, which is giving life beyond procurement, making it possible to conduct life-cycle assessment (LCA) (Harty and Laing 2010).

The early massing can also be tested for sustainable comparisons meaning that even at the early stages various options can be tried and tested leading to better-informed designs. Similarly, to sustainability, LCA is and will have a significant part to play in the procurement of buildings (Sørensen 2010). This is even more so, when the initial planning and post operations and maintenance issues are added. Suffice to say that best practice currently has three models running concurrently, one for the strategic policy makers or investors, one for the designers and procurers and finally one for the operations and maintenance people who pick up the pieces after practical completion. This is not optimal or efficient at all, because repeated input of data increases the likelihood of error which encourages a 'knowledge-drop' at each point of the saw tooth knowledge acquirement diagram.

Better-informed designs are possible, by bringing all stakeholders on board sooner in the process than previously. But while this is a bonus, it is also potentially problematic. Not least is how this collaboration is managed. While there is clearly a need for a manager, there is also a need for bells and whistles, with regard to authorship, quality and level of development, but this could well be dealt with using metadata, and machine learning capabilities.

The benefits of the model are not lost on some flexible entrepreneurs, already there are stakeholders who are entering into mutual agreements to work together to reap the rewards of completion on time and to budget. The biggest issue here is risk and how much or how well you trust your partner. Building trust in a business environment and especially in a fragmented market requires new skills and new procedures. Changing work practices from the adversarial to the collaborative requires major changes in mindsets and even social behaviours (Sigurðsson 2009).

Methods of integrating these diverse methods will improve how we make buildings and how we use them. Facilities management (FM) has a critical role to play here and methods of facilitating designers without alienating them will consume many resources before an acceptable solution can be found. The driving force will be collaboration and already we are beginning to see consortia being formed where certain players can work purposefully and profitably together to mutual gain (Smyth and Pryke 2006).

Transference of Information

The major obstacle to collaboration is how trust is nurtured, how new blood can enter the mix and finally how information, competences and knowledge is shared for the benefit of the team, the project and society at large. Initially, there were calls for sharing or giving away data for free, but with contractual obligations and recovery of costs there is a great reluctance to do so by the players and those who have invested so much into the project, to see others in the supply chain capitalise handsomely on their endeavours (Williams 2009).

How this can be remedied rests with the client and the appointment of all the stakeholders in the project. First, principles say that work effort must be remunerated, and second, there cannot be subsequent adversarial disputes about the quality and correctness of the data. The correctness of the data needs to be calibrated and one method is metadata (Onuma 2010).

A database is essentially an organised collection of information. It is usually digital (since the arrival of computers), which are programmable devices that can store, manipulate and present output in a useful manner. It is this number crunching effect that is of interest, in that there is potential for great computations. Finith Jernigan quotes Denis Diderot (1713–1784): 'There are three principle means of acquiring knowledge; observation of nature, reflection and experimentation. Observation collects fact; reflection combines them and experimentation verifies the result of that combination' (Jernigan 2007).

So here we have a link from Daniel Schön's thoughts on the reflective nature and the ability to combine and/or experiment and that link is the ability to handle data, which should open a whole new panacea to the designer. Kimon Onuma is one who has embraced this exciting situation with his 'BIMstorms' (a play on Brainstorms), where his motto is Keeping It Simple (Stupid) where data is taken and used (/reused) in various formats (Onuma 2010).

This is achievable using 'Cloud' technology. The term cloud relates back to telephony where up until the 1990s telephone companies offered point-to-point connections. Then came virtual private networks (VPNs), which utilised the latent bandwidth more effectively and opened up new possibilities. The term is derived from the cloud diagrams that were used to fence the differing tasks and entities that were grouped together, showing that they could interact on a platform or network level.

The subsequent persuasiveness of the internet brought computing into this mix, which was first noted in 1997 by Ramnath Chellappa (Wikipedia Contributors 2010), where data could be centralised and equally important access controlled. While all this may sound daunting and even tangential to the design process, Onuma makes one compelling argument. One aspect of design is about *programme* and this usually entails a long process of gathering information, setting out requirements and fulfilling demands. This can result in briefing documents of imposing size, (depending on the project) which need reading, interpreting and actioning. Often clients can have huge sets of demands, depending on branding, function or standards. The processing of this data can cause errors (human or whatever) through distribution and handling.

Unlocking Bottlenecks

Central to this process will be the expectations of the collaboration and this will be built on faith and hope (Smyth and Pryke 2006, 2008), giving rise to confidence and resulting in trust. The characteristics of trust were derived from the work of Lyons and Mehta, Smyth tells us, bringing the economic and social analysis of trust to relationship management. There are essentially two elements involved, the first is the self-interest part and the second is the socially orientated part, which demands certain obligations in a social network of relationships. This introduces reputation and advocacy into the mix and it is prolonged through experience.

Already in the United States, the National Building Information Modeling Standard (NBIMS) of America has deployed a compendium of principles called a capability maturity model (CMM) to facilitate this process, but where the National Institute of Building Sciences (NIBS) see problems is how to bring the strategists and the operations and maintenance people on board the grand coalition of consortia. Essentially, there are three pipelines and each has a different model and a different purpose (Harty and Laing 2010). But they all serve the same client and there is a need now for the client to step up to the plate and knock heads together.

Legislature has also a role to play here and as we have seen with sustainability, it can be done. Those authorities with clout, like the major agencies in the states such as the military or the state agencies in Norway who actually commission work and have a portfolio of properties to maintain are beginning to set demands which require a broad response possible only through consortia.

The strategic alliances made through these consortia will see like minds using the same tools to use, reuse and exchange data. There will be an acknowledgement of each stakeholder's worth and expectations for each stakeholder's input. The rewards will be significant and in proportion to each stake. There will also be continuity as the same methods and processes are honed and improved with each project. It will transcend procurement, involving the strategists and developers at start-up, the procurement team through construction and the facilities managers through operations until decommissioning and demolition.

CHAPTER 4

Education, Learning and Reskilling

Many companies succeed in offering loyalty reward programs to enhance repurchase rate, for example, the mileage of airlines (Liao et al. 2019). Liao et al. propose an architecture of Blockchain-based Cross-Organizational Integrated Platform (BCOIP) for issuing and redeeming reward points. BCOIP integrates a number of companies (i.e. countries/regions/authorities) into a consortium and issues electronic reward points, which are convenient to collect and can be circulated across companies. Also, the BCOIP system makes all the companies collaborate with equal rights and standing.

Ethereum is chosen as its backend. Ethereum features include decentralisation, openness and tamper-proof resistance. It is supported by the Ethereum Virtual Machine, with smart contract functionality, so that the system will be secure and transparent. It allows companies to supervise and maintain their platform and eliminates the need for central control. New blocks are validated by a consensus algorithm to become a record on the ledger. An Ethereum smart contract supports the operation of transactions without third parties and is written in

Transforming the Construction Industry with Blockchain: Enhancing Efficiency, Transparency, and Collaboration, First Edition. James Harty.

a high-level programming language such as solidity. It functions as a decentralised application, which are designed to run on a distributed computing system to avoid any single point of failure. DApps are stored on the Ethereum Blockchain, executed by the EVM and its smart contracts.

The web server is written by Node.js and Express, and web3.js is used to interact with the blockchain. The front-end webpage is written by HTML, CSS and JavaScript, and this works with the Ethereum blockchain, using Geth as Ethereum full node and MongoDB as a database to store all user account information and data. The functions include mint, deliver, redeem, takeover, issue and transfer, which will alter the state of reward points.

Currently, the BIMcert has two streams; trainee and trainer, this can be slightly modified into the learning phase (trainee) and also a register for work completed (trainer). The trainee engages in knowledge learning and skill-set acquiring, while the trainer displays the skill sets and demonstrates competences applied to new situations. This also gives the platform a circularity (life-cycle) aspect, showcasing good practices and rewarding better methods.

Blockchain offers the track and trace ability to log a person's progress through the platform. It can also certify their completion of modules and lastly it can reward that journey. While the platform is an authoring engine, the blockchain element makes it a recognition device. Completion of modules allows a coin to be claimed/given which accrues to a wallet, rewarding progress. It maps maturity and builds up a portal for users to promote their successes. This brings encouragement to the market too.

Work is needed here to define what is rewarded. One aspect is to reward the module completion and two is to reward uploaded work, similar to BIM360 or DALUX where incidents can be identified, remedied and signed-off. This requires an assessment of the finished work either if it complies then the coin is awarded or else some form of self-learning needs to be applied to ascertain the quality of the work.

Artificial Intelligence versus Human Intelligence

There was a discussion about how Facebook could be monetised (in 2007). The reality was that the 'like' button gave users the ability to gather basic information on their followers, but it gave Facebook even more: hundreds of thousands of new data points on each user's 'likes',

Fig. 4-1: Student refurbishment project (Tomas Bottenelli).

information that Facebook could compile and eventually turn into dollars (Kaiser 2019). At this time, Google was about information, while Facebook was all about connecting people.

For us, this monetisation means bringing manufacturers and producers into the fold. Bots allow us to see how many downloads, to whom and to where, meaning this can be leveraged to the supplier to follow-up on project-based modelling. So, you download and insert a VELUX roof window, VELUX can have a follow-up to either price it better, if it is not seeing it happening after tender. It also gives them feedback that their pricing was right or needs attention, and it allows them to see who got the bid, knowledge being king.

Allowing third parties into the platform means that we can also establish a job portal for our members. Students would be allowed to upload their CVs and portfolios for free or nominal charge, while employers could place adds for a fee, for a successful introduction (Figure 4.1).

Collecting data points requires some smart algorithms writers. Micro-targeting allows us to better match people with projects, both nationally, internationally and Europe-wide. Targeting is a finance source which can earn money, look at Zuckerberg.

AI Money versus Performance

I was first exposed to blockchain as an award in the construction industry using an AECcoin (Architecture, Engineering, Construction), but essentially in a passive mode, acting only as a verifiable credential.

Accumulating a collection of coins documented competences, beyond national boundaries on a Europe-wide basis. Next, I was an external advisor to BIMcert, which is a Moodle learning platform offering blended-learning modules to bridge the skills-gap across Europe in digital adoption, to improve the currency of skills on the market, offering smart readiness and a digital building passport.

This was poorly adopted and in a subsequent submission, for further Horizon-2020 funding, I wrote an improved strategy where the introduction of a reward system would change the engagement. Several outcomes became clear; rewards could be used to improve better practices; training paths could be devised for each user and that feedback could alert the platform which modules are being used and which might need to be reassessed, showing a prosumer approach.

Imagine three entities are bidding for a construction project. 'A' will always bid lowest, making up the difference in requests of information and change orders. 'B' has a good reputation, and if you wish to engage, it comes at a price. 'C' will give you the building you need for the next 20 years, saving a nominal 20% in energy and operational carbon. Currently 'A' normally gets the commission purely because of capital price. To get 'C' on to ballot-paper a new method is needed. It removes the Green-wash and rewards the savings made with a nominal 5% incentive to deliver the saving over 20 years. This is blockchain that either looks at the energy use through its life cycle and if the target is met, then the blockchain can automatically process payment.

Monitoring the interaction of the platform and user offers a new scenario and puts the platform on a different level, making it responsive and proactive in its engagement with the user. It also offers a monetary aspect where third parties can use the system to promote their own workforce competences, offer a skeletal structure to rollout their own modules and lastly offer a platform for accredited bodies to align their own institutional corporate membership exams. It can also learn from the users' interactions and build patterns to give feedback and reinforce a robust model offering improved outcomes.

Finally, after the platform's launch date, to remain relevant it needs a method so that new innovations can be incorporated on to the platform, by having both a trainee position and a trainer option, so that it can stay up-to-date in the future and be continually pertinent. The platform can also inveigle trends and identify tendencies that might not have been detected or thought of before.

During the development of this new platform, other things were advanced, namely that technology is moving at a rate of knots. First and foremost, new App's brought many options to the table. Open Badges became units of learning and in essence replaced CERTcoins, offered in the funding pitch. Second, shortly open badges will have blockchain capabilities. Third, that blockchain wallets will soon be open-sourced meaning that the technology is catching up with desired outcomes. Wallets will also allow for irrefutable verification of the credentials earned, making it desirable for users seeking proof of achievement and offering a verifiable source.

CHAPTER

5

Risk

> *'Imagine a world where all communications throughout the process are clear, concise, open, transparent, and trusting; where designers have full understanding of the ramifications of their decisions at the time the decisions are made; where facilities managers, end users, contractors and suppliers are all involved at the start of the design process; where processes are outcome driven and decisions are not made solely on first cost basis; where risk and reward are value-based, appropriately balanced among all team members over the life of a project; and where the profession delivers higher quality design that is sustainable and responsive. This is the future perfect vision of Integrated Practice'* (Broshar et al. 2006).

This was stated at the AIA conference on BIM in 2006. Thom Mayne was even more outright with his paper: 'Change or perish', if BIM was not embraced (Mayne 2006). Essentially, BIM was seen as a method to coordinate and collaborate, easing the adversarial culture that abounds within construction, and so reducing risk. Managing this new situation

Transforming the Construction Industry with Blockchain: Enhancing Efficiency, Transparency, and Collaboration, First Edition. James Harty.

requires automated procedures and robust structures to contain the possible outcomes.

Two reports from the 90s in the UK before the onset of BIM, set the scene, which began the reforms and mapped the process to change the construction industry (Harty 2012). The first, 'Constructing the Team' by Latham, recommended a need for better standards in construction contracts at a time when there was little or no cross platform uniformity. He called for better guidance on best practices and legislative changes towards arbitration in an attempt to change industry practices.

> '*. . . to increase efficiency and to replace the bureaucratic, wasteful, adversarial atmosphere prevalent in most construction projects at the time*' (Latham 1994).

The report wished to *delight* clients (Latham's words) by promoting openness, cooperation, trust, honesty, commitment and mutual understanding among team members.

Finally, he identified and determined that efficiencies, especially in savings of the order of 30%, were possible over 5 years. In the report, he condemned existing industry practices as *being* 'ineffective, adversarial, fragmented, incapable of delivering for its customers' and 'lacking respect for its employees'. Even at this early stage, he urged reform in the industry and advocated partnering and collaboration by construction companies. He went on to say:

> *'Partnering includes the concepts of teamwork between supplier and client, and of total continuous improvement. It requires openness between the parties, ready acceptance of new ideas, trust and perceived mutual benefit' and 'Partnering arrangements are also beneficial between firms' without becoming 'cosy' (sic).*

Of the recommendations in the report, the two most notable with reference to this research include:

> *'The use of co-ordinated project information should be a contractual requirement'.*
> *'The role and duties of project managers requires to be more clearly defined'.*

Following on from this seminal report came the Egan Report, 'Rethinking Construction' (Egan 1998), which identified five 'drivers' to improve in construction practices:

- Committed leadership.
- A focus on the customer

- Integrated processes and teams
- A quality driven agenda
- Commitment to people

A further four processes were identified where it was recommended that they should be significantly improved:

- Product development
- Project implementation
- Partnering the supply chain
- Production of components

Finally, there was a call for a set of targets to be improved with respective quantities for the improvements:

- Capital costs were to be reduced by 10%
- Construction time was to be reduced by 10%
- Predictability was to be increased by 20%
- Defects were to be reduced by 20%
- Accidents were to be reduced by 20%
- Productivity was to be increased by 10%
- Turnover and profits were to be increased by 10%

There was also a call for decent and safe working conditions, whilst improving management and supervisory skills. This was coupled together with long-term relationships based on clear measurement of

Fig. 5-1: Terminal 5, Heathrow, London, Richard Rogers.

Fig. 5-2: Student project (Erica Dima, Mathias Hammerdorf, Emma Grau, Gudrun Sebaldus & Nermin Al-Shakargi).

performance and sustained improvements in quality and efficiency. The best embodiment of these works and the culmination of Egan's work could be said to be encapsulated in the new Terminal 5 at Heathrow, opened on 27 March 2008 (Figures 5.1 and 5.2). Egan, chief executive of BAA, commissioned the terminal and implemented the 'Terminal 5 Agreement and the Delivery Team Handbook' (Haste 2002).

CHAPTER 6

BIM

Building information modelling has established itself in the market convincingly, and is immense in its ability to coordinate procurement, involve all stakeholders and digitalise the whole process to build. Most see modelling as a 3D exercise, giving a volumetric digital twin. But it is more, the fourth dimension introduces time, which means the building's timeline can be plotted. The fifth dimension means resources are brought into the equation and this grows to adding financial cost. Facilities management is regarded as the sixth dimension and sustainability is the seventh. Taking this to its conclusion, the eighth dimension is considered health and safety.

BIM's huge claim to fame is the ability to right-click an element and add properties. Simple as this sounds, it is the quintessence of the model. Adding properties means that data is being amassed and can be used to plot outcomes, test what/ifs and to reassure the projects validity. This brings reliability to the scene and allows all stakeholders to make meaningful and confident predictions, which improves learning outcomes and heralds better building practices.

Transforming the Construction Industry with Blockchain: Enhancing Efficiency, Transparency, and Collaboration, First Edition. James Harty.

At a basic level, it brings control to the project and allows for collaboration on the project. Construction is all about collaboration and this is why BIM has a home here. Collaboration requires trust and the model can allocate who did what and apportion ownership. This gets overlooked sometimes, but it needs to be stressed again and again. By allowing several stakeholders access to the model and giving them editing rights means it is a rich environment where all aspects can be tried and tested before a sod of turf is turned.

Admittingly, there can be unforeseen developments and issues not met before but this is the bell-ends of the spectrum. Eighty percent can be securely and confidently designed, allowing performance to make an entrance. This new platform then grows to be bigger than ever thought of before. The ability to test scenarios means that a better building is designed and planned as never before. Whether it is in the key junctions or renderings and fly-through, all aspects can be catered and delivered to all interested parties.

Ultimately, it will short circuit the whole plan-of-work as one-to-one modelling produces digital twins, which will alter work-stage payments and up-end the whole financing model. The financial bubble diagram, the building information model and the facilities management (FM) model will all intertwine into a single BIM pipe filtering the correct outcomes to respective stakeholders. These outcomes are immense when seen as the written word, but increasingly I am seeing relevance of it.

Currently, I have a 6-week elective, where the course consists of three parts; the renovation of a building project, the use of generative design to rack up the building parts and dynamo-code scripting to produce an element within the design. Once these have been produced, it is the intention that the metaverse be used to showcase the results and the exam will consist of a 3-minute video, documenting the design, the process, the detailing, with a voice-over and credit for the backing track.

This video will then be placed on the students' digital profiles and within days, the number of likes and invitations to internships grow astronomically. This is something the students fully understand and their attention to detail and storyboarding is something we do not have to teach; they get it and deliver it in spades. This is the crest of the wave for building information modelling at the Copenhagen School of Design and Technology's Architectural Technology and Construction Management course.

Collaborative Trust, Not Technology, Is Integral to This New Method of Working

When I first started embracing this digital world, I fully expected it to be the implementation of a new technology. What I was not prepared for was that the more immersive that modelling becomes the more one has to trust colleagues' work. This can be done through standards or by bridging skill gaps. The first can be administered through Information and Communication Technologies (ICTs), which are marshalled by a new role in the procurement team. The latter requires new methods to provide *just-in-time* modules, top-up post educational courses and by offering new accreditations, which are owned by professional bodies for their members.

Currently, this has manifest itself with ICT agreements which are the preserve of technology savvy stakeholders. The size of the task means that most limit themselves to BIM Level II because that was the threshold named back when this became mandated through governmental policy in the United Kingdom in 2016. In a real world, the more stakeholders involved the more data types that show up. Typically, this could mean the architect using AutoCAD or Revit, the engineer waiting for the architects to firm up their proposals, or in best case scenarios using Tekla, while the services might be in MagiCAD, and documentation in Microsoft Word, Excel or Adobe Acrobat (other software is found).

On top of this personnel might/will change, either within the companies involved or within the project team, and also the competencies of individuals on board. Levels of development also condition where and when decisions are made, with consequential impact. So, a complex matrix may arise from all these variables. The bottom line is trust and this is a human quality. Finith Jernigan addressed this in 'Big BIM little bim' (Jernigan 2007).

This has meant that the next phase of digital development has stagnated, and that BIM Level III is not happening at a rate that it should. One method of tackling this demise is to automate it. First, in this process, is to use cloud computing so that all data is held on a server where duplication can be sorted and minimised. This is controlled by common data environment (CDE), and usually when stakeholders are exposed to this it is a no-brainer in its adoption. CDE allows access to data through applications which can be on handheld devices instantaneously and robustly accurate.

Once the model is hosted on the platform, the virtual model can be mapped on to the reality either on site or in the refurbishment.

A mixed reality can be achieved where the model can be seen in its final position. If the fourth dimension (time) is also modelled, together with the fifth dimension (resources), then by going to today's date, the handheld device can show the model overlapped on to the reality. By traversing the site, one can find out if the project is before or ahead of schedule, costs can be tracked so that bottlenecks are avoided, placing all stakeholders back in control.

Tracking progress can also be captured by a laser scanning at regular intervals which allows the built environment to mapped back on to the model, where a clash-collision exercise reports that everything is as should be. Such a set-up allows for discrepancies to be highlighted so that the model or reality can be corrected in real time. This has the added benefit that the finished model is a digital twin, which is so useful for FM and lifecycle analysis post-procurement.

Beyond BIM Level II

BIM has a coordination role that is growing in stature as modelling becomes more centre stage and the driving force behind construction. The coordinator's role covers all tasks through the initiation phase through to handover and life-cycle analysis. They establish the BIM strategy involving the selection and the configuration of the tools to be implemented. Most importantly, the projects' coordinates need to be mapped (whether global or local), while selecting the data management system to be used. This defines the scope and extent to which the project will be identified. At a more intimate level, the structure of data is classified for all stakeholders, including naming conventions and nomenclature. At this level, drawing schedules, and things as basic as printing styles are outlined. All of this starts the process of data that is input and collected.

Next, the physical context is mapped, of the existing situation, including the topography, neighbouring buildings and infrastructure and their file types where compatibility is assessed. Increasingly, this overflows to geographic information systems (GIS) data which contributes to the existing situation, making it possible to easily map and rank all categories of information in a filtered routine that rationalises into streams of relevant data to all stakeholders in formats that the need and can access. The transfer of data and classifications is also relevant here too, as different silos want the data in different formats that should happen in a streamlined world.

This will be monitored and controlled by machine learning as the construction industry becomes more digitalised. It will roll over in the modelling management system deployed, where definition of the data exchange format will become less critical as it becomes automated. All users will get the information and knowledge that they need without having to translate and check the veracity of what is being pooled and shared. This will decrease the amount of written rules for the exchange of data and lessen the role of communications being monitored as currently established by BIM execution planning (BEP).

Also included in these decisions and role defining things comes the ambition of the project, the levels of development (LOD) and levels of information (LOI) as they become more relevant and intrinsic to the process. The platform becomes fundamental, whether it is cloud based or on local servers. Establishing logical naming conventions might be consigned to the past as naming becomes seamless to each and every user. We all know what water (UK), aqua (La), eau (Fr), wasser (D), vand (Dk), uisce (Ir), etc. can all be linked by the chemical formula of H_2O. Once this link is established, the world is your oyster as the potential and scope grows across the whole language sphere.

Being able to transform data, ensuring readability, make the legibility of data in differing formats transparent, meaning that many aspects of blockchain are being met and enacted upon without realising it in everyday encounters. Merging data and filtering it to specific requirements redefines how data is shared, and the custodians of that data become lesser managers of these flows. It means that Information Communication Technologies (ICTs) become visible in plain sight and do not control collaboration anymore. Collaboration becomes easier and effortless to many stakeholders as they need it or not.

BIM Collaboration Formats (BCFs) created to facilitate open communications take a backseat as this hurdle is removed for compliance as it is today. Scheduling, logistics and costings (Hardin 2009), instead of defining BIM managers, become automated and function outside their remit. This might/should open up new possibilities and potential to areas such as parametrical programming and coding. But even here there is capacity to trade and be aided by machine learning.

Generative design can also be mentioned here, and this can be part of the design process too. Design wizards can enforce consistency and standardisation. This in turn reduces legal and liability issues. A mechanism does not exist currently for rewarding designers to provide rich information models. This will add value to architectural services (Eastman et al. 2008).

Work stages and workflows will also change, and by definition contracts and negotiations will be transformed. Automated code checking customised to the location, being built, will make catalogue and type buildings easier to place. Lean construction will become easier to implement as waste can be reduced, removing errors and rework.

Designers and producers can better respond to clients' often changing needs. They can also better advice and coordinate all stakeholders input. The BEP provides a functional framework to ensure successful deployment of advanced design technologies. It is about optimising work instead of optimising siloed interests. It minimises surprises and redundancies in the flow of information. It can be dismantled by automating the process. Making the correct decisions becomes a machine learning exercise, which will only grow as the practice improves.

Many other procedures outlined in ISO 19650 and PAS 1192 frameworks become challenged as many of these actions become obsolete. ISO 19650 is largely formed out of the BS 1192. It addresses the whole life cycle of any built asset, including strategic planning, initial design, engineering, development, documentation and construction, day-to-day operation, maintenance, refurbishment, repair and end-of-life. PAS is a publicly available specification to develop standards, specification, codes of practice or guidelines. They were derived out of the UKs need to digitalise from back in 2016 when digitalisation became mandated for public procurements (Bew and Underwood 2010).

PAS 1192-2: 2013 deals with the construction phase, outlining the requirements for Level 2 maturity. It also sets outs the framework, roles and responsibilities for collaborative modelling. It also expands the scope of common data environments, which is patently needed currently. There is a focus too on project delivery because the whole digital workflow changes from many former fragmented practices. PAS 1192-3: 2014 deals with the operational phase focusing on use and maintenance of the asset information model (AIM) and FM.

PAS 1192-4: 2015 covers codes of practice, rather than specification standards, to document best practices for the implementation of Construction Operations Building Information Exchange (COBie). COBie is a non-proprietary data format to publish a subset of BIM into assets which are not dependent on proprietary software implementation. This allows data to be transferred to a bigger audience, as it is not software dependent. PAS 1192-5: 2015 frames security issues for the protection of the data generated for the built environment and smart asset management. This is a latent development that only arose when the value of data was realised. PAS 1192-6: 2017 addresses health and safety concerns, while PAS 1192-7: 2017 provides a specification to define, share and maintain structural digital construction product information.

These were all developed in preparation of the UK adopting digital construction in 2016. It is one of the few cases where a top-down approach is seen where legislation drives the adoption of BIM. Often there is a bottom-up approach, but usually, this is piecemeal and can develop unchecked, where no best practices are documented or promoted. Early adopters tend to be younger and have less experience and trust the technology to deliver the product. Older stakeholders tend to protect the experience that they have accumulated and do not wish to change something that serves them well.

Building Information Modelling (BIM) is not only an authoring tool for architects and engineers but also for all stakeholders and third parties in the building programme procurement process. Analysis tools like code checking of building regulations and environmental simulations that can report on heating loads, daylighting and carbon use will push the adoption of intelligent modelling faster and further than previously thought. Just as in the automobile industry performance will increasingly play a stronger role in the procurement process.

The benefits for clients should not be underestimated either and some are already reaping them where project certainty is to the fore. However, the professional language that architects and engineers espouse is a latent force that can run counter to fostering collaboration. Traditionally, they operated in their own silos and delivered in predefined formats in an exclusive protection mode. An emerging professional, the Architectural Technologist, can bridge that divide and adopt the adjunct role of manager in the integrated project delivery (IPD).

BIM unilateral adoption has been slow. There are a number of issues here and one is the entrenchment of the different professionals and their methodologies. While nobody challenges the right for an architect to control aesthetics and space, equally nobody questions that it is right for the engineer to control the structures and services. What is questionable is their mindset and language, if there is to be the real possibility of shared data, and genuine cross-discipline collaboration.

Sharing data and collaboration does not sit well with the disciplines involved in the building industry (Cicmil and Marshall 2005). Cicmil and Marshall elaborate and elucidate a scenario of pseudocollaboration, where a two-stage tender is hopelessly inadequate due to the intransience of the quantity surveyor (QS) in their perceived role of advisor to the client. This is akin to the difference, in the fashion industry, where there is prêt-à-porter and couture division serving two totally different audiences. One pays for the privilege, the other gets it as cheap as possible. In construction, there is no mechanism to allow the QS to enter into a collaborative state with the main contractor (and no desire to either).

Discouraging collaboration includes the treat of new entrants into the industry, the buying power of both suppliers and buyers, the rivalry among existing firms and the fear of substitutes, who might deregulate standards. These strong entrenched attitudes (Walker 2009) in the design construction divide were addressed in the procurement of Heathrow's Terminal Five (T5), delivered on time and to budget (Haste 2002), where such an environment was nurtured and encouraged (Ferroussat 2008). It was based on the principles specified in the Constructing the Team (Latham 1994) and Rethinking Construction (Egan 1998).

Had the British Airports Authority (BAA) followed a traditional approach T5 would have ended up opening 2 years late, being 40% over budget coupled with six fatalities; this was not an option for BAA (Potts 2009). By carefully defining responsibility, accountability and liability, the focus was firmly placed on delivery. Contract remuneration was based on reimbursable costs plus profit with a reward package for successful completion. This incentive plan encouraged exceptional performance with the focus on the issues of value and time. Value performance occurred primarily in the design phases and was measured by the value of the reward fund for each delivery team and calculated as the sum of the relevant delivery team budget less the total cost of the work of that delivery team.

BAA took out a single premium insurance policy for all suppliers, providing one insurance plan for the main risk. The policy covered construction and Professional Indemnity (Potts 2009). Sadly, while TS was collaborative it was not a virtually modelled project. Questions must be asked as to how much sway the various disciplines and the entrenched methods had in this change of mind. Or was the management chain of command too onerous. The team structure had a hierarchy of several layers of management; the development team, the project management team, delivery teams and task teams. There was no common model to reference and the level of comfort of the construction manager may not have been too cosy. Construction managers have the lowest level of comfort, working with other professionals (under 20%), while owners, architects and engineers have nearly twice that level (Eckblad et al. 2007), meaning that while the traditional demarcations have a good bonhomie, issues arise if the industry can afford this luxury anymore.

Back in 2008, there were many barriers to BIM adoption, that it was not needed, that existing CAD systems were adequate, that it was expensive, there was a lacking skill set, that it would not reduce drafting time, that it lacked features, was not requested by clients and also the other stakeholders (Tse et al. 2008). A year later the only one of these that still held water was the skill set (Economist 2009).

'BIM and Construction Management' by Hardin is a very practical oriented book (Hardin 2009). The sub-title 'Proven Tools, Methods and Workflows' is essentially, what it delivers. It presents an array of practical information, aimed at the user. It has tried and tested methods of binding the 3D geometry with the fourth dimension (time) together with the fifth dimension (resources), which can be assembled together in programmes such as NavisWorks. Here clashes and collisions as well as timeline monitoring can be mapped and resolved before becoming an on-site problem, as was the case traditionally.

Hardin discusses the parameters of management, preconstruction, construction, administration, sustainability and facilities management. Indeed, this is one of the first publications where FM is seen in practice as an intrinsic part of the building process (life cycle assessment), which now is a critical part of the whole process. The total integration is also well documented, where he shows that he is indeed on top of where the whole process is going. As can well be imagined, the integration of these differing technologies into the traditional work phases is new and can be daunting. Hardin gives step-by-step guidance in a straightforward manner.

The following is an example of the development of the procurement of buildings which are complex and invariably expensive as a result. It engages the route Frank Gehry revolutionised the procurement process and it high-lights how machine learning will drive the process even further.

Computers provide a means of building previously unbuildable works for architects like Frank Gehry (DIGITAL PROJECT – Frank Gehry.). He set up Gehry Technologies (GT) to realise his unique forms. Two sequential projects were the Walt Disney Concert Hall in Los Angeles and the Guggenheim Museum in Bilbao. With regard to the concert hall, Gehry found himself beset with cost overruns and the project was shelved for a period due to lack of funding. It finally cost an estimated $274m., which is more than five times the $50m. budget at the start of the job.

In this situation, Gehry has said that his position went from having the parental role at the start of the project where he was in control, to an infantile one when cost overruns threatened to scupper it. The focus moves from the architect to the contractor. The architect has lost face in the eyes of the owner and the contractor is now seen as the saviour if the building is to be realised. Conversely, when tendering came about for their next commission, Guggenheim Museum in Bilbao, GT sent a member of staff over to Bilbao to train the bidders in the software prior to tender, which was pretty unique then. The result was they came in under budget seeing more than a fifth being knocked of the estimate (Harty 2012).

How can one project with conventional tendering end up five times over budget while the other, using a common model, come nearly one-fifth under budget? The upshot is that subsequently people who wish to work with Gehry must adopt his processes and prequalify for collaborative work. It has put Gehry firmly back in the parental role at the helm of the ship and one where he is in control. It heralded a new dawn for Gehry, where he now uses selective tendering, and bidders learn how to extract quantities. The intelligent model (BIM) has done this for him.

From the evangelistic viewpoint, this is the clarion call, but from the practical position, there are many other issues. Primarily, there is ownership. Who will own the model, who will manage the model and who will co-ordinate the model's passage through its turbulent growth. In the Gehry case, it is a star architect and in such lofty situations, those choosing or succeeding to work with him have identified this type of work and accept its challenge. This can now be marshalled by bots and apps and work seamlessly under the radar.

The following is a dialogue with Frank Gehry from an exhibition presented in Copenhagen (Gehry 2008). Frank Gehry has said that the culture of architecture in our time works like this;

> *'You do a job; - you meet a client, they hire you to do a project, and it's usually a kind of a nice love affair and so on. It's a very positive, uplifting relationship at the start, and you develop a scheme, with plans for their building and they're upbeat and happy about it'.*
>
> *Of course, they have a budget, which they tell you and a time schedule or whatever. So, you finish the design and you put it out to bid, and then it comes in over budget. That (happens), I'd say, 80% of the time. Then the construction people say just that: we know what to do - straighten out a few things - we'll get it on budget.*
>
> *Of course, the owner finds himself very confused about this, for the most part, because they don't have the extra million dollars or whatever it is,*

and they're on the way or they're underway, and it's very hard to stop or be sympathetic to the architect, or to the project. They feel betrayed, and this happens all the time, and it's an uncomfortable place to be but no matter how much work you do, an architect can't control the marketplace, or the cost in the marketplace, or the construction world; you know, it's just not possible.

Now you can be as careful as possible about working for budgets, but I've always hated that moment, and 'my friends have always hated that moment and you sort of wonder is there some way out. In the middle ages, the architect was a master builder, they built the cathedrals, they were respected, they had a process, and it was done over centuries, so no one got the blame, (laughs). In our time, you have the Sydney Opera House where poor Jørn Utzon gets clobbered. It's a horrible story. It practically destroyed the man's life'.

So, in setting the scene, Gehry has recounted that a parental/infantile relationship occurs between the architect and contractor towards the client. Initially, the architect has the parental role with the client, advising and leading the way in this new adventure to build a house. After going out to tender the bid often comes back way over budget, and going cap-in-hand to the client a new price must be negotiated. At this point, the contractor is on board, he made the bid and on hearing the situation will usually offer ways of minimising the over-spend. This describes Design and Build (DB) scenarios over Design–Bid–Build (DBB) tendering.

He now takes on the mantle of parent and the helpless architect becomes infantilised, taking a back seat and losing control. Below, Gehry tells in more detail about the two commissions his firm had for two major clients. The earlier building is the Disney Hall in Los Angeles, which falls into the infantile arena. Here the tender sum came back sky high but luckily, in having a client as big as Disney the project was completed. But the new scenario is told in his next commission for the Guggenheim Museum in Bilbao where the technology restored the architect to the parental role. He continues:

'And so, on Bilbao, for the steel bidding, and there is not one piece of steel that's the same if you look at the steel frame, we used CATIA. We sent a team to Bilbao and spent a week training the Sub-contractors and those people bid on the construction the steel frame. They came in 18% under budget on just the steel alone. There were six bidders and the spread between them was 1%. Now that is knockout, rare, you don't ever get that. Which means, when you show a model of a building which looks

like Disney Hall to a contractor (which we did, way back), they give you a price that's out of this world. Until you say to them 'Here's a wall that we built with it, here's the drawings. Here's like how you can do it'. The guy said 'Oh, OK!' And then you get real. And that's what happened with Bilbao and that's what happened with all our projects since then. It's not that you control the market but that you can more precisely control the process and the things that can be controlled, you control, and it has worked beautifully'.

While on the one hand there has been an in-depth discussion of digitalisation and its impact on technology, there is an equally valid concern with regard to how management, and all it entails, embraces this virtual world. We have seen how the CAD era initially failed to bring management on board, and now how BIM is intrinsically not in the same process (to be renegotiated), but rather in a paradigm shift. So, how is the new message being brought across, applied and scaled?

BIM as Management Rather Than Modelling

Back in the mid-1990s, there was a marked period of innovation, in corporate IT, when enterprise software applications, typically Enterprise Resource Management (ERP), Customer Relationship Management (CRM) and Enterprise Content Management (ECM), became indispensable practical tools for business. Corporate investments in IT surged during this time from about $3,500 spent per worker in 1994 to about $8,000 in 2005, (according to the U.S Bureau of Economic Analysis) (McAfee and Brynjolfsson 2008).

With the economic collapse in 2008, Chip Jarnagin tells us that 'the failure rate for IT projects was reported to be around 70%, with the most cited reason for project failure being an unclear project vision/purpose as it relates to the business' (Basu and Jarnagin 2008). What is critical here is the how the message is interpreted in the business model. On the other hand, Jack Dorsey, the founder of 'Twitter', revealed recently that the mobile payments system, 'Square', of which he is CEO, is already processing an incredible $3 million in transactions per day. Just 3 months ago, it was seeing only $1 million in transactions per 24 hours (Essany 2011). Cleary there is volatility here, but we can also see that where there is clarity that the technology is winning out, especially if its potential is apparent and rewarding.

'In Praise of Dissimilarity', Gibbert introduces us to a completely different tack, also worth exploring. This involves the concept of 'similarities' and the association businesses place on them when evaluating capital expenditure. Managers' traditional understanding of 'similarity' has been taxonomic, whether explicitly or implicitly (Gibbert and Hoegl 2011). Taxonomy in general is the practice and science of classification.

So, when Intel, a computer hardware chip manufacturer, bought McAfee, an antivirus software manufacturer, for a record $7.7 Billion (their biggest in history). The Financial Times was very puzzled by the acquisition of these two most dissimilar entities. Apparently, in justification, experts envisaged that the chips can and will be improved against viral attacks. Nevertheless, cognitive psychology goes further suggesting that there is something called a 'thematic similarity'. Gibbert even supports this with a riddle; 'what has an athlete's footwear to do with a MP3 player, and the answer is in their association to working-out in a gym'.

So, in essence, there are new territories to explore and new markets to exploit. What Gibbert was telling us, was that there were new breeds of manager open to this association of dissimilar entities that would otherwise be blinkered using traditional methods. Finding these types in the construction world might be an awesome task, but likewise preparing for their appearance would be an astute one too. Where the argument begins to bear relevance is in his next comment that:

'The first camera phone appeared in 2001, but it took smart phone manufacturers six years to integrate the GPS and camera function'.

With regard to BIM, there is a similar situation. This can be found in the association of BIM with GIS. Essentially, they are two most dissimilar entities, with one based wholly in the 3D world and the other very much in data layers. Furthermore, there are not pan-uniform definitions of the term, and even more so, 'business model' terms. Mark Bew takes us through some very interesting definitions of BIM, including how it is interpreted by the various stakeholders, drawing particular attention to:

'. . . keeping critical design information in digital form' (to make it) 'easier to update and share, and more valuable to the firms creating and using it',

As well as:

'. . . creating real-time consistent relationships between digital design data, with innovative parametric building information modelling technology' (to make it) 'possible to save significant amounts of time and money and increase project productivity and quality'.

These two comments encapsulate very well much of what BIM was about to those who understood it, but as he rightly continued, they fall far short of what could or usually was presented to a CEO, or boardroom, in a language that made sense to them (Bew and Underwood 2010)

So, considering when Citibank decided to tactically force information technology on to electronic banking, it dramatically changed the banking. When American Hospital Supply decided to reshape its customers' supply chain and procurement processes through order entry terminals in customer sites, it positioned the company to compete effectively in healthcare cost management. When American Airlines pushed its travel solutions systems to customers, it changed the travel industry forever (Basu and Jarnagin 2008).

But despite these visionary examples, there is clearly a lack of functioning communication and interaction between general management and IT, which Jarnagin describes as a glass wall, not unlike the glass ceiling often alluded to in gender promotions in business.

He went on to define the cause for this:

> *'There are five primary reasons for the development of the wall: mindset differences between the management staff and the IT staff, language differences, social influences, the immaturity of IT governance, and the difficulty of managing rapidly changing technology'.*

The mindset was akin to left side (brain) people, largely IT personnel, talking to right-sided people, usually management. The person hired to bridge this gap was the ubiquitous Chief Information Officer (CIO), referred to here as the Chief 'Integration' Officer. The next was language and this was easily alluded to in the jargon that surrounded both spheres in their encoded acronyms. The next could be described as a cheap shot, but the social standing of 'geeks and nerds' within the IT community was easily recognised (rightly or wrongly), followed by the immaturity of this largely new discipline. Finally, there was reference towards the rapidly changing rate of the technology, and the resources needed to master it.

If the solution, as he said, was to 'detect, assess and respond,' then 'flexibility' was named as the key in healing this rift. This involved bringing in IT literacy into the boardroom and coupling this with effective leadership. This can be implemented with the notional rotation of management roles within these positions. It can also be helped by creating new demand, to align strategies across the board.

Finally, removing the jargon, rationalising the expenditure and creating an IT portfolio aided the demystifying of the two polarities, while bringing transparency and analysis, which ultimately restored confidence.

This warped thinking had progressed to the point that if you searched in New York for 'movie show times', the results now were tailored to that vicinity only, which of course, opened up innovative and opportunistic potential, lateral fields rather than closeted vertical silos. This implicit progression from a simple association has led to new markets and applications. Just as app.'s can now find a vacant parking space with a street or two of where you are searching, it is very indicative of how intuitive and indispensable that these small aides can become, and how apt they are becoming (Jarnagin and Slocum 2007). This is a good early example of automation making its mark in a parallel world.

This brought us back to the mindset and a new layer of thematic similarities in BIM. Once elements or components were formed, then data and fields could be attached to them. With the bridging of BIM and GIS, then a whole new plethora of associations, presented themselves. The Real Estate Norms (RENs) might link themselves to appraisal and brief work stages, giving a better-informed picture. Very soon, it would be a punitive exercise not to have these features included.

Isikdag draws attention to the single shared information backbone that BIM affords us. He also points out that some urban management tasks and cityscape visualisations are managed using Geospatial Information Systems (GIS). These two unrelated fields can be conjoined using a BIM web Service, which he calls 'RESTfull BIM'. REST is an anagram for representational state transfer, and this is a method of allowing both entities to integrate (Isikdag et al. 2010).

Dennis Shelden at a conference in 2006 spoke about sport stadia design and parametrics concerning the capacity and the breakdown, or the demographics, of the spectators. This was my first introduction to generative design. In real time, on screen, he entered a capacity for the stadium, together with a percentage for corporate hospitality, all the way down to the proportion and percentage of the cheapest seats, which resulted in a bowl design, meeting the specified criteria.

Changing any of the aforementioned fields met with an immediate response. Making changes to the form with sight lines or proximity to the playing area, reflected in the schedule reports generated from the model, informing the potential capacity and matchday revenue. He concluded that you would have to be suicidal not to implement this parametric before financing any sporting stadium (Shelden 2006).

What this reinforces is the notion that the model can be interrogated and analysed even at the formative stage of the design. The importance of this is slowly permeating through the decision-making process that engages management. This is even more relevant in Denmark today where carbon and thermal energy declarations need to be made at concept design so that these issues are part of the performance criteria to be incorporated into the design.

With regard to expertise in BIM, Randy Deutsch maintains that opting for depth over breadth is a false choice. He means that it will lead individuals, organisations, professionals and industry in the wrong direction. Expertise, he claims, is a much more social, fluid and iterative process than it used to be

> *'Being an expert is no longer about telling people what you know so much as understanding what questions to ask, who to ask, and applying knowledge flexibly and contextually to the specific situation at hand. Expertise has often been associated with teaching and mentoring. Today it's more concerned with learning than knowing less to do with continuing education and more with practicing and engaging in continuous education'.*

Developing this idea, he goes on to ask the critical question:

'The answer to 'should I be, a specialist or generalist', is 'yes' (both). There must be people who can see the details as well as those who can see the big picture. One gift of the design professional is the rare (and underappreciated) ability to do both simultaneously. As with any hybrid-generalising specialist or specialising generalist-one's strength provides the confidence to contribute openly from many vantage points and perspectives'.

He then draws on a metaphor in the shape of a capital 'T' to illustrate the two parts:

'It is critical for 'T-shaped' experts to reach out and make connections (the horizontal arm of the T) in all the areas they know little or nothing about from their base of technical competence (the vertical arm of the T). T-shaped experts have confidence because of their assurance that they know or do one thing well.' (Deutsch 2011)

This concludes with:

'Design professionals must be both BIM technologist and building technologist. Those who accept this model will lead, persevere, and flourish in our new economy'.

'It is not just that the integrated team is now multidisciplinary, but we each must become multidisciplinary. Doing so requires a multidisciplinary mindset. This entails empathy, a genuine appreciation for others' ideas, seeing from many perspectives, and anticipating possible consequences to any course of action'.

Before BIM, leadership and management was by and large top-down with someone senior designing or detailing, having some underling drawing it up. BIM demands a different workflow, namely side by side. Increasingly the senior and junior professional will need to use and help each other through the design process. However, it does not end there either, this binary operation grows into a network of others working together, and this network needs management and understanding.

With 'whole building design', the project team needs a collective vision. This is being heralded by Bill Reed as the 'Composite Master Builder'. The term harks back to the Master Builder of old. The intention is to bring all of the specialists together, allowing them to function as if they were one mind. The process avoids, as Mario Salvadori says, the 'reciprocal ignorance' of the specialists in the design and building field (Gabrielli 2010).

The team can include:

- *site professionals, such as planners, civil and environmental engineers, and landscape architects*
- *design team members such as programmers, architects, and interior designers.*
- *building systems experts, such as structural, mechanical, fire protection, and building science and performance engineers*
- *construction professionals, including cost estimators, project managers, trades people, and crafts people.*
- *owners, including financial managers, building users, and operations and maintenance staff.*
- *local code and fire officials*

An architect is ideally suited for the leadership of design teams, because of their legal obligations to the profession, comprehensive training in holistic problem solving, and an understanding of broad cultural concerns. This presumes that architects would maintain a clear overview of the project team's work, to oversee and coordinate the work of the project team.

The manner in which the design team is structured traditionally evolved around the architect as lead consultant. This role established the team, ran the job and administered the contract. The architect was the point of contact for the client and the architect often wined and dined clients, building up bonhomie and developing relationships. On large projects, those days are gone.

Design and Build, partnering, PPP and PFI all indicate that the architect's role has diminished, with managers, controllers and contractors now manning the pole position. A poignant clause in an AIA contract (B141/CMa) states:

> *'The Architect shall not have control over or charge of and shall not be responsible for construction means, methods, techniques, sequences or procedures, or for safety precautions'.*

Compare this to the personal architectural services of the legendry architect Frank Lloyd Wright, who wrote:

> *'The Architect undertakes to itemise mill work and material for the building, lets contracts for piece work and eliminates the general contractor where possible by sending a qualified apprentice of the Taliesin Fellowship at the proper time to take charge, do the necessary shopping and hold the whole building operation together, checking cost layouts, etc. . . and endeavour to bring the work to successful conclusion'.*

Non-architectural professionals might roll their eyes up to the heavens and ask why is the architect there with regard to the former quote, while lamenting the iron fist that Wright wreaked havoc with regard to the latter. I am not unduly concerned with both comments, but when Rab Bennets describes the damaging schism between concept and detail work stages, with the following, then there is something to answer:

> '*. . . there is a rapidly increasing tendency to separate the concept from the detail, a divorce that leads in precisely the opposite direction to architectural integrity*' (Bennetts 2010).

This chapter examines this development, looking at the procurement team, collaboration, trust and cost savings that can accrue from the emerging information technologies.

In setting out his stall, Patrick MacLeamy, CEO of HoK, has made two major contributions to the debate about how and why changes must occur (MacLeamy 2010a). The first is the now famous 'MacLeamy Curves'. This is a bell chart showing the expected resources plotted against time for a traditional project procurement process. Naturally enough, the biggest portion is tied up in design development, technical design and production information stages, beginning with appraisal and brief formulation and tailing off with mobilisation and practical completion. The procurement work stage is the meatiest and most resourceful stage currently in most projects.

With the adoption of BIM, more decisions and co-ordination mean that the bell chart moves more to the beginning of the timeline. Mapping the two charts against each other shows a telling proportion outside the original line and this is where the risk lies. Risk in this instance is that work is being done but not remunerated according to the work stages, meaning should the work stop or falter the design team is out of pocket.

With the hopeful normal running of a job, payment catches up in production where this slack is recovered. Balancing the books would suggest a reforming of the work stages to spread the risk and charge a higher amount for the new work associated with BIM. Some do this, but many architectural firms, while acknowledging this also point out that charging more earlier is not an option for many architect/client relationships. They cite that the client cannot raise funds to offset this spike in the process, as the financial institutes have their own guidelines and clearly, they have not embraced BIM yet.

The other contribution is 'BIM, BAM, BOOM', where appropriately enough BIM means Building Information Model, BAM; Building Assembly Model and BOOM; Building Operation Optimisation Model. MacLeamy portrays the benefits over time, in the following fashion:

> *'For every dollar spent in design, twenty dollars are spent in construction and sixty dollars are spent in operating the building over its useful life of fifty years or more'.*
>
> *'. . . BIM supports development and testing of design ideas. BIM also supports budget and programme compliance, Finally, budget enables compliance with energy goals'.*
>
> *'. . . Contractors do not build these days; they assemble manufactured products brought to the site. Contractors now use BIM as a Building Assembly Model or BAM. BAM allows better scheduling, BAM facilitates sub-contractor co-ordination, BAM supports cost control and BAM manages construction value twenty times the cost of design'.*
>
> *'. . . During its lifetime, an owner can leverage BIM and BAM to optimise building operation. I call the use of the model in this manor Building Operation Optimisation Model or BOOM. BOOM helps the owner*

manage energy consumption. BOOM also helps the owner with scheduled maintenance. Since BOOM is managing a value of sixty times the value of design, the cost savings potential is enormous'.

'... The real promise of BIM is better design, better construction and better operation. In short BIM, BAM, BOOM!'

These two vignettes are very compelling, the first addresses the evolution within the design team and the office and the second the lifetime analysis of the building or project. Taking the first on board could ultimately herald a production programme of zero fees in the distant future, due to automation. The second changes the whole design philosophy, as FM leverages its way into the design office with demands both for the founding principles of the process right the way through to scheduled maintenance. Architects must find ways of facilitating these two evolutions or risk being marginalised out of the process. In their defence, they claim that they are a holistic profession made for change management and new ways of working.

Examining the RIBA work stages and plan of work reveals many options in how they can accommodate differing projects and contracts, and much time and effort has gone into making these differing scenarios. A parallel development is the plan of work for the Office of Government Commerce 'Gateways' (OGC, UK), where the focus is to build up a programme of management, where the best policy makers have thought through the end-to-end process to translate policy into delivery plans and into desired outcomes (Crown 2009).

It is underpinned as such: strategic assessment: 'Is a programme (only) review that investigates the direction and planned outcomes . . ., together with the progress of its constituent projects. It is repeated over the life of the programme at key decision points'. The assessment is then coupled up with several reviews during the project as follows:

1. Business Justification: 'The first project review comes after the strategic business case has been prepared. It focuses on the project's business justification prior to the key decision on approval for development proposal'.
2. Procurement Strategy: 'This review investigates the outline business case and the delivery strategy before any formal approaches are made to prospective suppliers or delivery partners. It may be repeated in long or complex procurement situations'.
3. Delivery Strategy: 'This review investigates the outline business case and the delivery strategy before any formal approaches are

made to prospective suppliers or delivery partners. It may be repeated in long or complex procurement situations'.

(A) Design Brief and Concept Approval, (B) Detailed Design Approval followed by (C) an Investment Decision: 'This review investigates the full business case and the governance arrangements for the investment decision. The review takes place before a work order is place with a supplier and funding and resources committed. A project will normally go through on OGC gateway review 3. However, in some circumstances, it may be necessary for a project to repeat the review'.

4. Readiness for Service: 'This review focuses on the readiness of the organisation to go live with the necessary business changes, and the arrangements for management of the operational services'.
5. Benefits Evaluation: 'This review confirms that the desired benefits of the project are being achieved, and the business changes are operating smoothly. The review is repeated at regular intervals during the lifetime of the new service/facility'.

The philosophy of these reviews and checks is to ensure that government and estate spending is getting thorough inspection: Delivering value for money from third party spending delivering projects to time, quality and cost, realising benefits getting the best from government estate delivering sustainable procurement and sustainable operations on the government estate support the delivery of government policy goals improving central government capability in procurement, project and programme management and estates.

However, in true evolutionary terms, the OCG will be part of the new Efficiency and Reform Group (ERG) within the Cabinet Office from the 15 June 2011 (closing its website on the 1 October 2011). Paul Morrell, the (UK) Government's Chief Construction Advisor, essentially wearing the hat of the client or client advisor, in a 'BIM Roundtable Discussion' (Waterhouse et al. 2011) in the 'NBS BIM Research Report' contributed the following:

> '. . . *let's work off the same data, that should not be controversial. With the power of virtual modelling, it has always struck me as insane, that we find it so difficult to construct a 3D computer model, but that we go out there and do it in the wind and the rain at one-to-one, as an experiment instead, which leads to a massive amount of waste. So, let's build the model.*
>
> *It's almost literally true, that I retired from my day job (as a QS), because I did not want to sit through one more meeting, where we argued*

why a duct and a column wanted to be in the same place. We cannot begin to understand the cost of that, across the whole of construction . . .'

The roundtable discussion brought many key figures in the UK construction industry to clarify what BIM is going through collaboration, adoption and discussing the role of who manages BIM. Klaschka said BS 1192, (a standard for managing the production, distribution and quality of construction information, using a disciplined process for collaboration and a specified naming policy), manages the model quite adequately, meaning that you do not need a manager. He continues:

'. . . the architect is responsible for co-ordinating the zones in the building at design (stage). If the structural engineer's structure is in those zones, and if the service engineer's services are in those zones, then spaces that are left are big enough for the people to wander about in, and do the things that they want to, then the design of the front end of the building is there.

What BS 1192 sets out, is the way for that to work. One of the debates that is going on in the industry at the moment is should you adopt a more risky approach than that, and I think that this is where the term BIM Manager and management of the model comes from. At the end of the day the architect is responsible for coordinating the stuff that is in the building . . . that is what is in our contract'.

In reply to this statement, the chair asked the table '. . . is it the architect, who manages . . .' which was met with a resounding 'no'. Morrell even interjected:

'that the idea of the model being managed by BS 1192 is fascinating. (continuing, assertively that) There will be managers! There is no doubt about it . . . If you are looking for a career for your kids, a model manager would be a pretty sustainable profession!'

Clearly, the difference between the architect and his colleagues was deafening. He bravely and courageously made the first gambit, but the other stakeholders were having none of it. Collard contributed that:

'. . . we (Laing O'Rourke) have got more traction in the areas of the business where we have employed BIM managers who invest in the training and take people along the journey. That is not just our own internal company, that is allowing designers to engage in the technology, to hold their hand, (and) to talk about how you coordinate a model. These people are

doing it day-in, day-out. They are problems that they are now able to fix, (typically) how we integrate models . . .'

Instead of integrating models by pressing buttons and drawing them in, 'we are now doing it automatically. That is saving hundreds of hours. I think a BIM manager is essential. I think that documents such as BS 1192 are fantastic, but what we need to realise is that it is all about behaviour, re-enforcement, and that writing a protocol, putting a document in place, is not going to guarantee that you are going to get the output (required)'.

Klaschka attempted to reiterate his position, only for Morrell to interrupt:

'. . . you describe the BIM manager; you said that the architect takes all responsibility to co-ordinate the work of all the others. That would be a terrific world, I would love to live in it, and if it happened consistently, then the architect would be a BIM manager'.

The chair then alluded to this being an opportunity, and that where opportunities like this happened in the past, new professions appeared. For the chair, being an RIBA delegate, this was significant, and I will return to this statement. The debate then shifted to a new area, where Kell commented that in his experience, it was generally the lowest tender that got the job. Morrell again, took up the cudgel:

'. . . because the industry has no better proposition!'

He would prefer to move away from the 'lowest competitive tender' to 'show me your best school' (for example). He elaborated that there was a contractor in the States that was offering 'zero-change-order' contracts to clients, meaning that if you do not change your mind about the brief that the tendered price is your contract sum. He added that, as a client, this is a very potent/desirable offer.

Øivind Rooth, then Deputy Director General of National Office of Building Technology and Administration (http: //www.be.no) from Norway, gave a keynote address about how they are dealing with the matter at the building Smart Conference in Copenhagen 2010. Norway is a country of 4.7m people with a Gross Domestic Product (GPD) of €60,792 per capita, meaning they are relatively rich (Rooth 2010). This means that they are good at investing in infrastructure and one such project is dealing with energy and the amount of energy buildings consume.

The parliament has recently had a white paper on building policy and has gone on the offensive. It has looked at their building stock and found out what are the desired qualities for their lifestyle in a cold climate. This has led to the introduction of legislation and now they are identifying qualified practices and practitioners to deliver their goal. This leads to better products and most importantly better processes as Rooth says himself.

He then outlined what his government can contribute:

- Predictable framework conditions for the players
- Development of good, coordinated instruments.
- Deliberate use of model projects
- Requirements beyond the minimum for all state-building and all tenancies
- New certification schemes
- New contract forms

His agency, BE (www.BE.no) had then acted as an instrument providing a knowledge database, regulatory incentives, tools and methods and appropriate information. This culminated with 'Byggsok', which was a public system for electronic services in zoning, building and construction matters. Its vision was to improve productivity and efficiency in the planning and legal process. Its objective was to facilitate this process through the internet.

The next phase was to go directly from design to permit, using code-checking, (an automated process on an open-source BIM platform). This was achieved by using and reusing data in the planning and building process. First, there was the data exchange with the project and the central server. Data was then added from the national registries, and this was sent to the municipality where the approval was processed. It then updated the national registries again. General, site specific (or local plans) and building regulations were all checked.

The checking relied on two, up until now, disparate technologies. One was BIM and the other was GIS. Construction information was extracted from BIM, while geographical information came from GIS, together with zoning and property information. Both were based on open-source platforms with BIM relying on IFC, and GIS relying on Geography Markup Language (GML), an ISO standard developed by Open Geospatial Consortium (OGC). The programme has not been rolled out yet, but in tests, there is a 20-minute turnaround (instead of the 3-month building permit normally experienced. Open standards have been insisted upon so as not to be dependent on or tied to proprietary systems. In addition, the major issue in amalgamating the two systems was that one is largely vector based, while the other is raster based. The various layers have grown out of two totally different systems that were never intended to meet.

The system is fully operational but has never been launched. Ironically, opening platform would result in redundancies in the planning departments and trade unions resisted its implementation. So, if the pressure is not coming from within, then what will drive the changes? Clients were instrumental in the DWG format being adopted as deliverables when CAD established itself, and they appear in the factors influencing BIM as having 49% influence. Code Checking's appearance was at 25% in the McGraw-Hill Report on Interoperability which was significant, in that there was not widespread checking then, so it must be determined as a 'wish-list' item (Young et al. 2007).

Tomas Pazlar (Pazlar and Turk 2008) found that moving a simple wall in and out several programmes led to data being dropped. Typically, a field would have no corresponding field in the new format and if not critical would be dropped. On passing back that field would be voided. Even using IFCs, evidence was shown that all export functions were not supported. It could be as innocent as the wall hatch or pattern being lost in a vertical section, but even so it meant that the operator had to be vigilant 'not blindly trusting the mapping process'. Machine Learning corrects this.

Alan Baikie of Graphisoft argues in Building Design's 2008 World Architecture 100, an annual survey of the top architectural firms in the world, that larger firms are slower to invest heavily in newer technologies in terms of money, time and effort in their migration into the 3D realm, leaving the door open for nimbler firms (Littlefield 2008).

This is what can be done now, augmented reality is beginning to emerge in the construction industry. Augmented reality superimposes virtual data about your surroundings via a mobile phone. This layering can be filtered to your requirements. Suppose you point your mobile phone at a historic building, the in-built GIS knows where you are and which direction you are facing, and so can recognise the building in question. Historical facts can pop up, details can be displayed, information about the architect, style or previous residents can be solicited. This happens now in museums where information can be given for each exhibit, in your preferred language. Point your phone at the Abbey Road pedestrian crossing and see John, Paul, George and Ringo superimposed on the street.

Taking this further, point your phone at a building site and see what the building will look like when finished. Give it to a sub-contractor and he can see the installation he is required to do. Give it to a construction manager and he can see if the progress of the building is up to schedule and if the reality matches the model. Give it to a surveyor and he can see the position of all utilities in the ground, no more hitting water mains when excavating.

Replace the phone with spectacles and a raft of information is available on the inner surface of your lenses. This is happening with the visors of fighter pilots and the dashboard of some concept cars where the focal point is set to infinity. This mapping of the reality with the virtual world completes the call for adopting the model. The closest you can get to a real building, that is not a real building, is a virtual one, and BIM is that situation. Before a real building exists, it has to be modelled in some matter or form (Pazlar and Turk 2008). Common Data Environments are providing these features today.

There were other drivers too initially, and prime among them was the momentum of modelling, the pervasive use of IT in general, especially in construction, and the need to perform better. Modelling is showing us that there is a new style of architecture, where forms previously impossible are now appearing on the skyline. IT has quietly evolved and is having a huge impact throughout the world. As cloud computing asserts itself, not being in the (cloud) loop will be detrimental to how you perform in the supply chain. The interweaving of BIM and GIS will also bring untold benefits and overlaid with Apps and Bots will reap a harvest of new opportunities beyond procurement and well into life cycle analysis. This is changing at a rate of knots and will only escalate with artificial intelligence.

Design strategies and management structures are also changing. BIM with better clash detection, better communication and better co-ordination is resolving problems earlier meaning that fee scales will soon follow the money and reduce the risk. This means that more will be done in the earlier stages and ultimately production information will be fully automated. Management is moving into relationship roles rather than previous hierarchical ones and this is seeing more side-by-side situations where people work together instead of vertically in the same organisation and across disciplines in order to do the job correctly.

This means the construction team is undergoing a transformation too. The right person for the job is improving. The design team is growing with everyone from client and the legal team, all the way to contractors and sub-contractors now having a say. The procurement process is expanding to include the financial model and the life cycle model, meaning the practical completion handover is not the final goal anymore but rather the start of the building's life, in which all stakeholders have an interest and duty to consider. Blockchain will lead and drive the reward system to implement these developments.

Architectural technology is also changing, instead of tried and tested details and constructions being applied and modified through time and experience, simulation and analysis now allow for the testing to happen digitally. As stated elsewhere:

> *'We should build prototypes in the inexpensive virtual world, not the very expensive real world'.*

Soon 'wear and tear' will also be another simulation to be applied, water penetration, and frost action will soon be performed on the model, and the results analysed, reducing thermal bridging, purely because it is analysable. Climatic data can be applied to the site location, the building's forms and compared. This brings confidence and certainty to the project. Novel technologies can be modelled in place and tested to achieve better performance, rewarding the entrepreneur rather than penalising the untested.

The architectural technologists' role is growing too. With their all-round ability and awareness of the other disciplines within the industry, their role is the glue, which keeps everything together. Whether BIM stays as a single entity or a federation of models, to be monitored and controlled, the role of the BIM Manager is certain. As Paul Morrell said emphatically in 2011:

> *'There will be a BIM Manager'.*

This corroborated, Jonassen who had championed for someone to manage the sharing, integrating, tracking and maintaining data sets, which requires overall leadership as far back as 2005.

Bringing Performance into the Design

It is one thing to make a 3D model, it is quite another to use it for 4D, 5D, 6D and 7D. Briefly, 3D is the X-, Y- and Z-axis of the model, but it includes the existing model, scanning and topography, it includes health and safety, renderings walk/fly through, prefabrication and laser driven field layout. 4D is the fourth dimension, time, so this includes phasing simulations, Lean scheduling, Just-in-time planning and visual verification for payment approval.

5D covers resources, and this is primarily costs and estimates, which includes real-time cost planning, quantity extractions, trade delivery verifications, value engineering, prefabrication, mechanical, electrical and plumbing and unique architectural and engineering elements. 6D has been assigned to sustainability covering conceptual energy analysis, energy analysis, sustainable tracking and embodied carbon assessment.

Taking this system further 7D has been assigned to FM applications, including Life-Cycle Assessment (LCA), Life-Cycle Costs (LCC), As-builts, Digital Twins, COBie data extraction, maintenance manuals, sensoring (monitoring dashboards) and end-of-life disassembly and transformation. There can be arguments for 8D is Health and Safety, 9D Lean Construction and 10D industrialisation but there is little support for them as it takes the system to extremes, which can be argued as being covered in the first 7Ds.

Paramount is that all of the above essentially opens the can of worms to address performance, because now we are testing the robustness and value of the solutions chosen in the life of the building beyond procurement or handover. Currently, both the design team and contractors see delivery as closure to a project and are built to leave it and move on to the next project. Many forms of contract have been developed to extend that relationship but even when entrapped into a life-cycle contract some players will go as far as selling their risk on to third-party hedge funds who take on the risk and manage the situation.

Trying to find the lowest common denominator for any continuity through all of the above it is the materials, that can hold all the properties to be found to drive the show. It is akin to elements in chemistry as being the lowest common denominator in science. Because of the climate crisis, where construction accounts for up to 40% of fossil fuel consumption, carbon has become the major driver having an influence across the board.

Materials

Mentioning a material to a handworker implicitly implies the scope of work to be done, by whom, for how long and at what cost (Palsbo and Harty 2013). Having a right-click function in BIM allows this to be embedded into modelling too, reducing the risk of human error. It also means that after sketching a building part, a schedule and an estimate can be given with all the information needed for a specification.

Costs begin with a unit price of a material, which combined with a cost factor where the total cost can be shown including labour, material cost, leasing of equipment, peripheral costs, fees and a cost index to relate the work to where and when it is to be performed. There is also an exponential value so that the if volume of work increases the overall cost decreases.

Other properties include u-values, fire-ratings, acoustics and embodied carbon. Because it can be automated it is there for reuse, resulting in time and cost savings in future projects. It can be as precise as naming suppliers and products. This can be described as a quantitative material library. Materials are rarely used in their raw form, usually being a component that has a performance to live up to. These include loadbearing capacity, climatic envelope, insulating qualities, fire ratings, moisture control, ventilation, wind performance, maintenance and acoustics. Others could include aesthetics, cost, quality, ease of manufacture and erection (ibid).

In the past, a single material could have several purposes or functions. For example, a brick could be all of the above, but not the best. Modern constructions tend to be sandwich components where each element has a particular purpose, whether it is loadbearing or waterproofing, and the technique is to place them together with appropriate tolerances, to allow them to cohabitate in harmony together.

Given that the scope of BIM is to gather as much relevant information in one place as possible so that as many stakeholders as possible can access it, makes it possible for all to have the latest information (Eastman et al. 2008). Effectively sharing data makes the model the primary tool for document generation to calculate, simulate and analyse. Patrick MacLeamy in his time against effort presentation sees this as moving the effort over time to earlier in the project's timeline, resulting in much better building and much happier clients (MacLeamy 2010b). It has the ability to impact costs and performance, track the cost of design changes while reinforcing BIM workflows.

Properties can be customised and added to the modelling software. In order that this process does not become too big and difficult to navigate, templates can be made to filtrate the workflows and bring some purpose to the process. Schedules can be generated from the modelling software, and these can be populated with price books to add time and cost to the work. This can then be used to formulate specifications both for the works and the trades. The material library is the most basic element to use for collecting this data.

CHAPTER 7

Performance

Designer Reputation

When the act of 'drawing' became what can only be called formalised (whose growth can be said to have blossomed during the Renaissance), there developed a separation between the drawing and its procurement. This meant that the architect was no longer the master builder on site becoming a third party for the construction and being removed from site, creating a studio. This is not to say that they did not visit sites, but they were not permanently there. Recently, David Ross Scheer, in his book 'The Death of Drawing, Architecture in the Age of Simulation' wrote:

> *'. . . whereas architectural drawings exist to represent construction, architectural simulations exist to anticipate building performance'.*

This applies a new deeper task on the media, it is no longer passive but gains an active dimension in its purpose. Meanwhile, Paolo Belardi, in his work 'Why Architects Still Draw' likens a drawing to an acorn, where he says:

Transforming the Construction Industry with Blockchain: Enhancing Efficiency, Transparency, and Collaboration, First Edition. James Harty.

'It is the paradox of the acorn: a project emerges from a drawing – even from a sketch, rough and inchoate – just as an oak tree emerges from an acorn'.

He tells us that Giorgio Vasari would work late at night 'seeking to solve the problems of perspective' and he makes a passionate plea that this reflective process allowed the concept to evolve, grow and/or develop. However, without belittling Belardi, the virtual model now needs this self-same treatment where it is nurtured, coaxed and encouraged to be the inchoate blueprint of the resultant oak tree. The model now too can embrace the creative process going through the first phase of preparation, where it focuses on the problem. The manipulation of the available material can then be incubated so that it is reasoned and generates feedback. This serves to align this shift in perception, methodologies and assess whether the 2D paper abstraction still has a purpose and role in today's digital world!

Increasingly, we are seeing digital handheld devices making an impact on building sites. The use of QR codes allows relevant data to be accessed *just-now*, where and when it is needed. The ability to check up-to-date information means that errors can be identified immediately and assigned to be corrected by the person finding it, all the way through to the person in a position to correct it. The single point of reference means that the red-lining or issue noted can be cleared and assigned a new stature as being fixed, all in the cloud, meaning the speed and efficiency of such methods become quickly adopted because they are seen to work.

Moreover, the stakeholders can be both 40/40 people. This generally means that those over 40 years old and often with the experience but challenged by the technology can engage and effortlessly make calls. While those under 40, who are tech savvy and app crazy, can learn and progress in the real-life jungle of the building site, where anything goes. The ease of use has been paramount in the adoption of these methods and is reaping rewards in less litigation and a keenness to complete the job properly.

Other stakeholders, notably contractors and sub-contractors are also engaging in such practices, which not long ago would be deemed highly unlikely. Sisk, a major contractor in Ireland recently completed a Center Parcs development at Longford Forest Holiday Village. The site is 164 Hz consisting of a sub-tropical swimming paradise, aqua sauna building, arrivals lodge, cycle centre woodlands buildings, pancake house and associated amenities and facilities. Accommodation consists of 470 lodges (5 types) dispersed into 78 clusters plus an apartment building bringing the provision of 500 units in the development.

BIM was not part of the project requirements, but Sisk sought to table the process at tender and use a digital project delivery (DPD) approach which was accepted. Center Parcs has now been implemented in all their developments. Synchronisation was achieved using drone footage, which was weekly, and mapped through Synchro so that critical paths could be monitored and maintained. The project was fast-tracked with a value of €65M and it took only 68 weeks in total. They were able to produce three lodges a day using prefabrication techniques and through DPD and lean construction, they implemented 4D construction sequencing, building information modelling (BIM), ASTA progress mobile app and field view (Kennedy 2019). Field view allowed for quality sign-offs using handheld devices without the need for paper work or returning to the site office.

Field view saved an hour a day, for all those involved, allowed 18,500 planned inspections, saving an average 15 minutes per inspection, resulting in 4,625-man hours saved in total. ASTA is planning software that allows capturing the work-rate on site, with reduced manual input and paperwork, which increased awareness through the team, with quicker data entry and increased added value for planning.

With regard to lean construction, three methods were deployed, work sampling so that the assembly was tested as being the best method of construction so that all were on board, just-in-time delivery and kitting, which means each job is quantified at source in the storage area so that everything needed to complete a task went out with correct quantities, resulting in no waste, and a more efficient work flow. There was a 22% productivity improvement observed on key trades.

Respect

Reputation is merely an opinion held about someone who is respected or admired, based on the past behaviour or character. Designer reputation is derived from the opinion gained through good design, whether it is aesthetic, functional or whatever. Within architecture, there is a term *starchitect*, blending star status with architectural status. This applies to architects such as Norman Foster, Richard Rogers and Zaha Hadid, among many more, but this reputation is squarely in the aesthetic field.

For many, drawings are synonymous with architecture, whether they are objects of beauty or technical efficacy. Architects and everyone involved in the design team, know and appreciate their worth, and all disciplines invest time and effort in their respect of the medium. This is painstakingly acknowledged in David Ross Scheer's book 'The Death of Drawing, Architecture in the Age of Simulation' (Scheer 2014).

That said, he goes on to say:

'This long tradition of drawing in architecture, with its influence on the thinking of architects and on the very nature of architecture, is in question for the first time since the Renaissance. Whereas architectural drawings exist to represent construction, architectural simulations exist to anticipate building performance'.

What he is alluding to here, is that change is upon us and that a paradigm shift is happening. He is stating that simulation is predicting building performance. Whereas aesthetics used to be all-consuming in the best architecture, a rider or a qualified subset is raising its head. Performance is being mandated by thermal comfort environments, by climate change and not least by diminishing fossil fuels (Harty and Miller 2014). It brings a balance and appropriateness to the equation so that we are 'able to sustain' (i.e. sustain-able).

This means that there is 'a new kid on the block', where increasingly, drawings are being overlooked, with the BIM taking centre stage and becoming the modus operandi. At a micro-level, the most obvious example is where Integrated Concurrent Engineering (ICE) is making inroads to design team meetings and on-site. ICE is lean conceptual design, where it can determine value parameters that are useful; update parameter values immediately; transfer parameter values automatically; link parameters seamlessly; perform analysis while, all-the-time, checking accuracy and applicability of the analysis, as it is performed (Coffee 2006).

In practice, it amounts to all design-team participants making their models available to each other, in a confederated model, in whatever medium, so that each's model can be superimposed against all others, meaning there is a concurrent real-time model available for the meeting. With each participant able to see the implications of their efforts overlaid on the others' work, faster solutions can be found, and moreover, immediate decisions can be taken, and not banked for discussion back at the ranch, as it were, meaning the solution does not need to be first tabled at the next design meeting, with all the repercussions associated with that process.

'It is an approach, fostering dynamic communication and cross checking, tightly integrated tools with their users and with each other, and focusing participants on the critical information pathways feeding design deliverables' (Coffee 2006).

As a methodology, it allows tasks to be conducted in parallel rather than serially, forming a critical path, whereas previously each discipline would release drawing sets in a round-robin fashion, for them to be coordinated locally. Normally, this would lead to clashes and conflicts being raised in subsequent design team meetings, if not on site, for resolution. This loop is now removed.

We are also seeing the emergence of design command rooms, which foster an environment where collaboration thrives. Central to a command room is a digital white board and integrated computers where co-ordination can occur. They allow real-time visual communication, for all stakeholders, creating a milieu to encourage collaboration. In such a situation, the white board can have a physical dimension of up to 4 m width but can be up to 90 m in virtual width. This means panning is possible using a hand motion and that several agendas can be housed at once from modelling to excel spreadsheets, Gantt charts, etc., which can be addressed using markers or post-it notes, where the content is then digitalised on the hosting computer in real time. Finger movements can be used to expand dialogue boxes and notes can be added too.

Different stakeholders' models can be superimposed over each other, and clash detections and best fit can be made in real time, at the meeting, rather than being tabled in a further 14 days, or whenever the next meeting is scheduled, as in traditional execution. Meetings can also be hosted in-absentia or remotely, meaning that it can be scheduled where each stakeholder attends via an online meeting or video conferencing. Avatars can allow attendees to seem to be gathered together, giving the impression of a cohort being together in the same room. It all happens in real-time and is better known as ICE.

David Miller, of David Miller Architects states:

> *'At the moment design decisions are all about reputation. A design that is seen as 'good' will enhance your reputation as a designer. In the future designs will be measured against performance. And that performance has a very direct effect on the financial reward you can expect from good design, from commissioning a good designer'* (Malleson 2016).

Naturally, everyone does not adopt the new systems and technologies at a uniform rate. New and recent BIM adopters will have to go through a managed process of change in their internal organisations as well as the external processes. They will also have to reconsider how they interface with the supply chains, clients and consultants (Johansen 2015).

Because of these variations, differing levels have been defined to inform and help users assess the value and quality of the information being offered and shared (NBS 2015). Level 0, as it suggests is at best

2D draughting. Output sadly is paper based without collaboration. Level 1, moving forward, is digital with 2D computer-aided design (CAD), often with 3D conceptual embellishments, usually in the form of renderings and perspectives. Sharing of data happens here to a degree, but mostly in finished formats, thus lacking cross discipline collaboration in its truest sense.

Level 2 is mandated in the procurement model being sought in the UK for all public buildings from 2016 (Waterhouse et al. 2011). Collaboration is required here, but it takes the form of federated models only, so that information can be shared. ICT contracts are needed to address issues of copyright and liability, and to allay fears and offer robust appointment documents (Waterhouse et al. 2011). These protocols are governed by processes described in PAS 1192-2: 2013 and PAS 1192-3: 2014. The first is a specification for information management for the capital/delivery phase of construction projects using BIM. The latter controls the operational phase of building assets using BIM but parsed from objects to assets, meaning they can populate a spread sheet and be useful to users, owners and facilities managers (FM) (NBS 2015). PAS means a publically available specification, which essentially fast tracts standards, specifications, codes of practice or guidelines developed by sponsoring organisations to meet market needs (Designing Buildings Wiki 2019).

Using the Level 0 to produce 2D drawings, with a lack of coordination can be said to increases costs by 25% through waste and rework. Between Level 1 and 2, 2D and 3D have a better probability of removing errors and reducing waste by up to 50%. Under Level 3, it is then possible to reduce risk throughout the process and to increase the profit by +2% through a collaborative process (Calvert 2013).

While the majority of the BIM users are still working in the Level 1 process, the more experienced users are seeing significant benefits by moving up to Level 2 (Bew, Underwood 2010). It also shows that it is important to improve competences and try to reach Level 2 before the majority does, in order to gain a market advantage.

> *'It is clear that organisations adopting BIM now will be those most likely to capitalise on this advantage as the market improves'* (Johansen 2015).

It all begins with BIM; the architect uses 3D modelling to investigate options and test building performance early on in order to optimise the building's design. The design is then handed-off to the contractor who streamlines the building process with building assembly modelling

(BAM). This allows for a significant decrease in construction costs. Once complete, BAM is turned over to the owner and becomes building owner operator model (BOOM). This allows the owner to manage the building over time and ensure optimised building performance throughout its entire life cycle. The real promise of 'BIM-BAM-BOOM!' is 'better design, better construction (and) better operation' (MacLeamy 2010c).

The term 'BIM-BAM-BOOM' explains how we should address and think about the model throughout a project. Whereas BIM is related to the building's information and the design's development, BAM is related to integrated project delivery (IPD) in the construction management, where BOOM is in the facility management phase (Harty 2012). Increasingly, large public clients are mandating the use of BIM in open formats, to have a better overall picture.

Thomas Johansen writes, from his experience during his internship in Oslo, that;

> *'Meanwhile, we are placing ever-greater demands on our built environments, and by adding multiple components to our buildings, we increase the risk and probability of errors. The ability to just delete or change something in the model is much easier than with paper drawings. The model also allows better control over buildability, progress, access to solutions and project economy'.*

Typically, a project manager might hold fortnightly project review meetings. This might occur on Thursday, even weeks. This would require delivering updated models Tuesdays, even weeks. BIM coordinators then would assemble federated models and begin collision reports to be ready for the review.

ICE is this relatively new design management system that has had the opportunity to mature in recent years, to become a well-defined systems approach towards optimising engineering design cycles. It encourages an idea that all elements of a product's lifecycle can be taken into careful consideration in the early design phases. Secondly, the concept is that the preceding design activities can all be occurring concurrently. This includes establishing user requirements, propagating early conceptual designs, running computational models, creating physical prototypes and eventually manufacturing the product.

In practice, this can mean all technical stakeholders at a design meeting might be wielding handheld devices as they, in real time, address issues normally noted and taken home from meetings to be rectified by in-house staff and presented at the next meeting.

Beyond the design team meeting, the debate was about 'if, why and how', the model might be given to contractors. Today contractors are demanding models in any way, shape or form, as it provides some sort of flow across the process. Needlessly, the better the model, the better the flow. If there are high demands for facilities management information, and if there is not, why not, then the debate now progresses to how to deliver the model to the client/user. Here there is a translation need from objects to assets. Paper abstractions do not, in any way, assist this process (Johansen 2015).

Reward (Incentivisation)

Traditional construction contracts, generally, are Design–Bid–Build (DBB) where the lowest or most preferential tender is awarded the contract. This has been hailed as flawed for many reasons, topmost being that in order to secure the work, the bid usually has to be the *lowest*, and if below cost, that the only remedy to increased revenues comes from shortfalls within the contract, the documentation or the drawings. These are litigiously examined and result in requests for information (RFIs) rework, delays and poor workmanship (Egan 1998).

RFIs usually seek to clarify further information or to provide information that was not complete at the time of signing the contract. It is good practice to include in this information, the affected parties, dates, any supporting documentation, as it will form a chain of information, creating a matrix to be tracked, answered and distributed appropriately. If this process constitutes a variation, it might qualify the relevant party to an extension of time, or a claim against losses or expenses, delaying the completion date and budget if not carefully managed. It becomes a phenomenon in itself (Aibinu et al. 2018).

BIM can mitigate this process for the better, and various contract forms strive to address these deficiencies. They range from Design and Build (DB) to IPDs and lately; insurance-backed alliancing (IBA) where the whole team makes joint decisions from day one for the benefit of the project (Thompson 2019). This involves the insurers keeping checks and balances in the form of three new roles. These are the financial independent risk assurer (FIRA), the technical independent risk assurer (TIRA) and an independent facilitator (IF). The first checks the cost plan, while the next assesses feasibilities, while the last stops the team reverting to type with adversarial behaviours. Blockchain could make tremendous inroads to this process.

Its intellectual position could be said to align with the Latham report (Latham 1994). This report wished to delight clients (Latham's words) by promoting openness, co-operation, trust, honesty, commitment and mutual understanding among team members (Harty 2012). Incredibly all of these aspirations have remained on the agenda right up until and including today. Finally, he identified and determined that efficiencies, especially in savings of the order of 30%, were possible over 5 years.

In the report, he condemned existing industry practices as being 'ineffective, adversarial, fragmented, incapable of delivering for its customers' and 'lacking respect for its employees'. Even at this early stage, he urged reform in the industry and advocated partnering and collaboration by construction companies. He went on to say:

> *'Partnering includes the concepts of teamwork between supplier and client, and of total continuous improvement. It requires openness between the parties, ready acceptance of new ideas, trust and perceived mutual benefit' and 'Partnering arrangements are also beneficial between firms' without becoming 'cosy' (sic)'.*

Many would say that BIM is delivering this nirvana and that collaboration is growing and benefitting better construction, but only to a degree, I would argue. There is no commitment to occupation, circular economies or sustainability, especially with regard to carbon. Handover is still seen as the end of the contract and in this all responsibilities vapourise shortly after the building is finished and the snagging complete. Increasingly, we are seeing owners/occupiers not taking over the facility but leasing it back from the builder, ensuring that the risk remains with the contractor, safeguarding any building malfunctions, whilst they are in residence. While this is a tendency, it is not the solution either. A better method is needed to deliver the environment we need and deserve. A better method is needed to encourage those providing this environment to perform. And finally, a better method is needed to reward such practices and promote working together.

Internet of Things (IoT)

The Internet of Things (IoT) describes a network of physical objects that have sensors, software and other technologies embedded in them to autonomously connect and exchange data with each other. It has become widespread, due to the number and prevalence of digital devices, namely mobile phones, handheld devices and the coverage of internet penetration across the globe.

Even RFIDs embedded in prefabricated construction elements can respond to facilities management operations and other maintenance manoeuvres. Cisco Systems estimate that the ratio of people to things grew from 0.08 in 2003 to 1.84 in 2010. This was predicted to grow to 6.58 by 2020 (Evans 2011).

It gives rise to smart homes, smart cities and a plethora of new interconnected networks, which can predict, track, be proactive, or prescriptive and optimise operations. This also means that it can be a big technology disrupter, as it makes analogue methods redundant. In security, it can protect investments and research against hacking or attack. It can offer Software as a Service (SaaS), offering cloud-based solutions.

Edge computing is a process where data is filtered down into quantifiable bites before being transferred, meaning that the process is leaner, resulting in cost savings and increased sophistication. Data analytics adds value to the data collection and this artificial intelligence (AI) is playing an ever more important and influential role.

Artificial Intelligence

AI is gaining enormous traction recently. After being broached first over 50 years ago, it went into hibernation, becoming unfashionable and derided even (Metz 2021). It engages neural networks, which alludes to being an analogy between human learning and machine learning (Rich 1985). Essentially, learning is about making associations between vast amounts of information, finding and recognising patterns in the presented data. This covers images, texts and translations but requires massive input to start the learning process. This is difficult to square with human learning, given that a child can make intelligent decisions with a smattering of information.

So, to paraphrase, the machine learning part requires numerous images of cats, for example, to be able to identify cats or even generate an image of a cat. Generative adversarial networks (GANs) relied on a first neural network to build a network, then another network to learn from the first one. This meant that one was trying to fool the second into thinking what it produced was a real image and this is critical in the learning process. This is the beginning of the process of learning because this is what it learned to do.

In making AlphaGo, a neural network could play and beat humans at Go, a game. Go is an abstract strategy board game for two players in which the aim is to surround more territory than the opponent. The game was invented in China more than 2,500 years ago and is believed to be the oldest board game continuously played to the present day. It is a grid of crossing lines (19 × 19) where the two players place black and white stones on the intersections to surround territory, the winner gaining the largest territory. Where chess has 35 possible moves per play, Go has 200 (Metz 2021).

So, just as IBM's Deep Blue beat Gary Kasparov in 1997, AlphaGo beat first Lee Sedol in 2015 with an audience twice that of Superbowl. Later in China AlphaGo beat Ke Jie in 2017 where the audience was bated to reach millions of Chinese, only for it to be pulled just before the start, as the match was being used by Google to rekindle relationships with China (having been excluded for over 7 years over research engine manipulation). This second match was a walkover because AlphaGo had been playing itself in the intervening years learning all possible moves.

The other main area of impact was text translation as in Google's translation service. It broke sentences into pieces, converted them into slivers of another language and then worked to connect these fragments into one coherent whole, giving rise to awful translations like:

The spirit is strong, but the flesh is weak.
to (English to Russian to English again)
The vodka is good, but the meat is rotten.

The new method was to do it in a single learned task where suddenly things went from incomprehensible to comprehensible. With the help of a new Google computer chip that could translate what previously took 10 seconds could be done in milliseconds now.

When humans learn something difficult, they usually build on related sources of things that they do know. For a computer to do this apparent random network is difficult, and when Marvin Minsky wrote a book called 'Perceptrons' in 1969, he claimed XOR functions were beyond machines. XOR is exclusive Boolean logic (two plus two is four), and this began the exodus in this area of computing. It was not until Geoff Hinton in 2012 showed that a neural network could recognise common objects. He established a company DNNresearch with two students in Toronto, which Google acquired shortly afterwards for $44 million.

Google's spend would rise astronomically in the next few years, DeepMind costing $650m, installing $130m on graphic chips alone, and as high as $1.2 billion by 2020. Peter Lee, the vice president of

Microsoft, is quoted in Bloomberg Businessweek as saying the cost of acquiring an AI researcher was akin to the cost of acquiring an NFL quarterback. Amazon, Facebook, Baidu, and finally Microsoft all started throwing incredible amounts of money to adopt this new technology to their brands.

Bots (AKA Robots)

Augmented realities are virtual worlds that echo reality, agents are catalysts that help the operation. Ontologies are the classifications that result from the whole operation. Basically, what Agent Augmented Ontologies do is number crunching, to find acceptable solutions for the various voids in the design. These results put names and products against the performances set out or demanded from the specification. There is a rapid growth in applications to analyse functionality in the early stages of design. As generic forms are tested for their constructability, methods are required to swap-in the components as they are prepared without jeopardising or compromising the remainder of the model until completely transformed.

They come in two types, proactive making the groundwork and reactive confirming best practices. Proactive means that the bot will do the data crunching and offer results based on the criteria desired, this could be cost, time or life-cycle issues in proportions specified by the client. Reactive means that the work will still be done but decisions will be left to the designer. This means that as each object is identified the bot will find products and assemblies that meet the requirements. The next step is will it act on the findings or stash the results in a property drop-down box for later selection.

This can be best explained using LODs, at 100, objects are generic and not specified, at 200, the bot will be finding products meeting the criteria, at 300, the design team are happy to go to tender, and at 350, the appointed contractor orders the product and starts a supply chain. Now at each stage, the question is whether the bot makes the decision or human input controls the process.

Typically, this might mean that a generic window is placed into the model. During the design phases, this window will undergo many operations to get the design right, including tests for daylighting, ventilation, energy frame and even fire escape, all before the process begins of finding a manufacturer. At this point, each manufacturer's window must be popped into the model for testing, until either a winner is declared or an acceptable range of products is found to comply.

This same method can be applied to all elements and components of a building in a new and very exciting manner. Previously, several somewhat subjective procedures had to be executed, totally divorced from geometric model or drawings. These involved listing all requirements in a tabular form with remedies and compliance notes or sources to the demands. What followed next required sorting them under importance and finally ranking them with weightings to particular instances. The best performer then got picked. The problem with this method is that is divorced from the model in a parallel process, needing to be synchronised and applied to both tracks, which is not optimal and could be described as double-work.

This is currently a technologist's job, to find appropriate solutions for the job in hand, backed up by a pseudo-scientific model to demonstrate documentable procedures which allowed closure at each work stage. This removes the possibility of renegotiation at later stages. This is not said negatively, but rather as an acceptable method to make logically sound decisions. The problem with the system is that it is removed from the model and as such relies on the technologist's experience and know-how, again it must be stressed this is not a bad position. In fact, many of my colleagues would hold that this is a core competence for a technologist. The biggest issue here is who is experienced and how long and through what processes this competence is achieved.

But imagine a scenario, where Obonyo describes a buyer's agent who needs to match a designer's requirements together with the manufactured products found in brochure (Obonyo 2010). Built into the search process there are building regulations to adhere to as well as cost, materials and any other peripheral criteria that might go towards making the optimal decision. The agent assembles the products that meet the conditions set out. The designer makes a choice, and the subsequent documentation is then drawn up.

Various cycles are performed to achieve this outcome. The various products are identified and compared. The relevant data is extracted and compiled into legible form. A secondary round might now take place refining the first set of results with more detailed data. The distillate surfaces again. The actions that have been performed include widespread communication to identify products that meet the desired standard, this is gathered from manufacturers. Next, an internal process takes place, to weight the selection set and rank them in order. The final communication phase suggests the preferred options, which are presented for nomination. Once the choice is made, a secondary process compiles the specification documentation in a useable form for the project.

The object of the exercise is to extract the performance requirements from the designer and turn this into a detail product specification. The method has been to interpret the designer's requirements and to make the result useable. Imagine no human involvement except at the user interface. These are called 'Bots', as in robots, and they are simply software applications that run automated tasks that are relatively simple and structurally repetitive.

A natural consequence of this is who gets selected and who is omitted. As a result of this good performers and poor performers become identified. If patterns appear, the effects become manifest. This allows a poor performer to assess their methods, pricing and quality, while better performers might get lazy or nonchalant. How these issues are rectified offers a new paradigm, which has not been experienced yet, but if faced might result in other concerns abounding. Taxonomies will rank, rate and report on remaining relevant.

In a sense, this is what Google search engines are getting better at doing, and it happens instantly or in the background. This automates the demand process and soon it will be linked or embedded in the model. Already type codes or instance codes can be attached to the building elements or components. By this, it could, for example, be a brick or the whole wall composition. A code can be as simple and as passive a blank field into which a piece of text can be placed either as a marker or as the link, that ties the process together.

Soon it can also be an active 'bot' that once released can go about its business assembling data for better selection. Its scope can be curtailed to generic choices before tender or in other forms of procurement, it can go direct to the manufacturers. Once the decisions are taken, analysis cycles can test the candidates for best fit, allowing the designer to select the best suited.

Imagine a bot placed on a door going off quietly and finding all the doors that meet the requirements demanded for that door. Initially, it might only be an internal single leaf door, with 23 manufacturers that fit the bill, but by the time, it is fully commissioned it might have a gained fire rating, sub-master key with particular hinges, a particular type of wooden veneer, a specific model and price, with a specified life expectancy with inbuilt inspection periods or repair schedules. This data is only relevant to those who need it but at each stage of the process, there is fingertip informed data at the ready awaiting selection.

Finally, bots will be able to harvest data, report how many objects are inserted into the model, do they remain throughout the design process and if not offer sales-teams ammunition to revitalise their methods. With the arrival of machine learning, the potential grows even more. Once a building is laser scanned, bots, or what are called agents, can begin to identify 3D geometry. Once the objects are classified, they can be named, costed and qualitied.

Moreover, this same procedure can be implemented on a building's point cloud, walls, floors and ceilings can be automatically recognised, and this short-circuits the whole transfer process in converting point-clouds to 3D objects. This greatly reduces the work involved currently and automates with certainty the geometry captured. As a starting point, this reduces the margin of area that can escalate with an unreliable first capture. This argument alone brings digitalisation to the table as a no-brainer, with the laggards and paper and pencil brigade, who firmly believe that *if it ain't broke don't fix it*.

Smart Objects

Herlev Hospital, in Copenhagen, was commenced in 1965 and opened in 1976, but the procurement process involved the design team getting a sign-off from the surgeons and medical staff in the early work stages for the operation theatres. Drawings, specifications and budgets were presented and agreed, but when handover happened, the staff found that nothing was where it was needed, things were missing and that a total refit was required, double work.

This is a classic example of double work where the language of drawings and specifications is not legible and readable for (lay) people (outside the design profession). In this instance, it was only when the property was occupied that the issues arose and needed to be addressed post-occupancy. Machine learning is changing this scenario enormously. As mentioned earlier, a laser scanner can capture several typical theatres (previous examples or examples of good practice). AI will then start identifying the 3d point-cloud into a 3D mass first, and then will start a process of learning the form of the differing objects and begin a process of breaking-up the model into its composite parts.

This process can continue down to all furnishings and equipment being identified through all items from surgical lighting devices to sharp scalpels and life-monitoring machinery. As each element is identified, it is removed from the scene and placed into a library. After several instances, the machine learning will improve and learn more as more scans are logged. When the sign-off comes, the users can be invited into multi-reality worlds (artificial reality, virtual reality and mixed reality) allowing avatars to inter-react and, in an immersive world, the theatre can be planned with accuracy and precisely where each and everything is needed. This space can be revisited and is always up to date, meaning the whole process has been bagged, ready for use.

AI can also allow an empty room to be furnished and laid-out as would best suit the purpose required and this can be approved by the users and clients before any decisions are made. This stream lines the whole process and the surface is still only being scratched. The more machine learning that happens, the better the possibilities. Agents can moderate and control the solution, but the verification can be supplied by blockchain. Exactly how this happens is not totally in place, but it is not beyond the realms of possibility, given all that has happened recently. Machine learning will be a mega-player in the new construction era.

Large Language Models

As the name suggest, large language models (LLMs) are big, enabled by accelerators that process large amounts of data. The enablers are agents that have no precise goal, other than to collect data. But as agents, they are intelligent and can think-out loud. It has what is called a Describe, Explain, Plan and Select (DEPS) method followed by a reflection mode that learns over several repeated occurrences.

They use intelligent agents to perceive environments, make independent decisions to achieve targets, that may/should improve performance, through machine learning. Therefore, this is the vessel that takes its inventory on-board and processes this data into usable code. Once assembled it can be retested, presented as *de facto* or be part of a new deep search to take it to the next level.

The difference here is the rate at which the learning occurs set against the human model. Humans rear children over many years and the bond has many passionate or slow-learning techniques. Both parents commit long term to this process, whereas machine learning casts itself into the problem and classifies the results, *no love lost*. This means that the exponential learning curve is worrying to where it can go and what it can achieve.

Morality has also raised its head in the debate, but here blockchain can play a role so that the process can be kept in check. Machine learning does not have a moral compass, and many will tell of incidents where fake truths, lies and deepfakes are perpetuated, challenging the validity of the practice. But if blockchain is the single source of truth, it will out and be victorious. It is hard to find academic evidence of this due to the time limit currently on AI as most of the tasty bites have happened very recently. But given the two polarities, blockchain will win-out.

Taxonomies (EU Directive)

This directive has six objectives to navigate the transition to a low-carbon, resilient and resource-efficient economy. They are to mitigate climate change, adaption to climate change, to sustain and protect water and marine resources, to transit to a circular economy, to prevent and control pollution and to protect and restore biodiversity and ecosystems (Slevin et al. 2020). It wishes to make a significant contribution to these objectives, while not impacting any in favour of another, by doing-no-significant-harm (DNSH). They impose minimum safeguards to help all, and it will help grow low-carbon sectors while decarbonising high-carbon ones.

Very few sectors of the economy are operating at a net-zero level, and emissions are not reducing fast enough, in general. The EU's Action Plan on Financing Sustainable Growth called for the creation of a classification system for sustainable activities or a taxonomy. In May 2018, the European Commission issued a proposal for a regulation which sets out the obligations for investors and the overarching framework for the taxonomy. This will be supplemented by delegated acts containing the technical screening criteria.

The taxonomy identifies three groups of users; financial market participants, large companies who are required to provide non-financial reporting and member states setting standards. This will ensure that it is rolled out with an impact offering a green deal or sorts. It will have a trickle-down effect that makes it adoption across the EU.

CHAPTER 8

Better Practices

Coen van Oostrom, a realtor in the Netherlands, wanted to develop the world's most sustainable building, according to the BREEAM rating method (van Oostrom 2016). BREEAM measures sustainable value in a series of categories, ranging from; energy, health and well-being, innovation, land use, materials, management, pollution, transport, waste and water (BREEAM 2019). Previously, he claims, there was a building in London, which reached a 96% rating.

His client, a well-known firm; Deloitte and the building The Edge, is situated in Amsterdam. It has over 32,000 sensors measuring occupancy, lighting, temperature and air quality and is connected to your smartphone via an app so that it can see who is where and when. This means that they could reduce 4,000 workspaces to 2,000, based on employees' schedules, ranging from sitting desk, standing desk, work booth, meeting room, balcony seat or concentration room. It also knows their preference for light and temperature, and it tweaks the environment accordingly (Randall 2015).

Transforming the Construction Industry with Blockchain: Enhancing Efficiency, Transparency, and Collaboration, First Edition. James Harty.

They achieved 95% through their methods, and upon inquiry found out that to get the extra credits that something *innovative* was needed. They talked to Siemens and General Electric, but it was Phillips who had a new system called *'Light over the Ethernet'* or PoE (*'Power over the Ethernet'*, now), which did not use normal power cabling but rather the ethernet (Philips 2019). The significance of this is that while the cable supplied power to the fitting, it became a two-way street, which also allowed the fitting to be monitored, (when it was on, for how long, etc.) to meet sustainable goals for optimisation and productivity.

At the end of the TED Talk, van Oostrom pointed towards smart cities and talked about being off-grid and sharing excess energy, but without mentioning how. Blockchain could have provided a structure to implement this. Brooklyn Microgrid does just that (Brooklyn Energy 2019).

This chapter essentially plots the impact and reach that blockchain will have across the construction sector, by imposing digital methods and means on to all stakeholders. While not wishing to sound condescending, it makes sense and sets up the grounds for adoption. It also promotes digital sharing and handheld points of delivery on the site.

Jason Farnell stated that the ratio of risk to reward in contracting is seriously imbalanced or that the margin contractors expect to earn from projects is insufficient for the uncertainty and risk exposures they face in delivering (Farnell 2018). Carillion one of the UK's largest contractors collapsed with £1.5B in debts caused by underbidding contracts with low margins. The industry expects 2–5% profit margins, the reality is closer to 1.5–2% (Wearden 2018).

According to the Construction Financial Management Association the average, pre-tax net profit for general contractors, construction profit margin is between 1.4 and 2.4% and for subcontractors between 2.2 and 3.5% (CFMA 2011).

More recently, many would see this as optimistic, as some feel that it is closer to half a percent, as in Construction News, which went as far as to say:

> *'In its 2017 analysis of the results of the top 10 largest contractors, including Balfour Beatty, Interserve and Laing O'Rourke, Construction News found that they had an average pre-tax profit margin of −0.5% as losses from problem contracts took their toll. Pre-tax losses totalled £52.9m on a combined turnover of £32bn. Meanwhile, the website theconstructionindex.co.uk reported profits for the largest 100 firms of just £1.1bn, equivalent to a margin of 1.5% on a £69.1bn turnover'* (Galaev 2018).

The collapse of Carillion, one of the biggest corporate collapses of modern times in the UK, has been attributed to an aggressive approach to risk transfer. PFI, or *'Private Finance Initiative'*, contracts were introduced in 1992 in Britain. They are used to contract out public services – anything from building a hospital to providing school meals. PFI has been criticised as being exorbitantly expensive, and also for hiding the government's true financial liabilities (Wearden 2018).

Wembley Stadium (affixed sum contract) was finally finished and handed over in March 2007, by Multiplex for the sum of £798M, rising to a staggering £962M (Scott 2006). It was late and long over budget (£445M – James Report). Multiplex made significant losses and tried recouping money by suing Mott MacDonald for £253M, Cleveland Bridge £38M with numerous other claims against sub-contractors and consultants.

Clearly, a new approach is needed. One which looks beyond mere procurement but embraces life cycles and circularity. There is a line where handover is seen as the end of the contract (snagging aside), and where there is no incentive to provide the facility, which is best suited or needed, with little or no further involvement.

Current methods address a bespoke system where it is more *couture* rather than *prêt-a-porter*. Admittedly each building is unique and no two are the same, even if it is only its siting. The tendering process while trying to encourage best value more often than not resorts to the cheapest bid. Tendering parties, knowing this to be the case often go in with a *loss-leader* to win the bid, reverting to litigation to extract further funding during the construction process. A *loss-leader* is a bid which makes no-profit (at outset), in order to win the business, with the intention to find faults, requests for information and change orders, that will open the possibility to get extra work, extensions of contract and generally generate new work to gain better remuneration.

Alternately, the cultural change required to implement integrated practice delivery (IPD) is an enormous challenge defining *'true partners and collaborators with a mutual interest in a successful outcome'* (Smith and Tardif 2009). Essentially, it alters the way and amount of time consumed in being adversarial and in expecting litigation. Increasingly, contracts are explicitly saying that stakeholders will not sue each other, that future legal action is a no-value task and that trust with verification mechanisms will become standard, as in banking. The principal cause of a bank failure is often a loss of trust rather than insolvency, there is very little difference between a failed bank and a health one, Smith tells us.

This is most evident in Information and Communications Technology (ICT) Agreements. Publicly Available Specifications (PAS) are fast-track-standards, specifications, codes-of-practice or guidelines developed by sponsoring organisations to meet an immediate market need. They are prepared following guidelines set out by British Standards Institution (BSI) (Designing Buildings Wiki 2019). In general, they create a milieu where there are methods of addressing who does what, who corrects mistakes, and who gets paid for the extra work, and how hand-over of information and data is conducted and classified.

Cloud solutions allow all data to be placed in one place, reducing the opportunity of error, due to numerous, conflicting, out-of-date information. Protocols exist here to make standards and expectations from all stakeholders. They can be hosted, and programmes and software allow the ease-of-use, especially for the older non-technological personnel who play an important contribution but can be technically challenged, when presented with new alien methods from what they are used to dealing with in previous projects.

Different packages offer different solutions but the most popular include BIM 360, Dalux and Solibri, without being exhaustive. Each stakeholder uploads their contractual commitments, and then these can be superimposed, assessed against other work, to eliminate errors and co-ordinate the design at an early stage. Moving on-site all can access data through handheld devices, meaning that data is up-to-date and relevant for whoever is accessing it. Tasks can be initiated here too and tracked to completion, which improves the paper trail and time previously experienced.

How this impacts technology is principally in the transfer of information and the risk it imposes on the authoring party who could be held responsible for the quality, completeness and accuracy of the handed over data. If a 'no-fault' policy is in place each stakeholder accepts the data as 'found' and must validate it, appropriately to their needs. Validation consists of two parts, determining if the data is from a trusted source and confirming the integrity of the information itself. Smith calls this stewardship. Where there are errors or omissions methods will have to be effective to compensate the corrector or rectifier instead of identifying the responsible party or assigning blame. The blame-culture stagnates the process and causes delays. There has to be a hand-off of responsible control.

This greater dependence of stakeholders on each other can cause strain within the working relationship if trust is not present and more importantly earned. In order to minimise and in an endeavour to make the process more transparent standards are invariably required. The National Building Information Modelling Standard (NBIMS) of America has deployed a compendium of principles called a capability maturity model (CMM) to define goals and offer methods of measuring business relationships, enterprise workflow, project delivery methods, staff skill and training and the design process against an index (Smith and Tardif 2009).

This allows for a form of benchmarking and acts as a quality management control for all those involved. It covers the data's richness, life cycle views, roles or disciplines, business processes, change management, delivery methods, time lines response, graphical information spatial compatibility, information accuracy and interoperability support. However, it is only a skeleton, which can offer the stakeholders an index to measure or check each other out, and to bolster their own pitch by giving them the tools to build their own argument and set out their own stall.

Data has value that the owner wishes to maximise should it be exchanged by recovering the costs of it and increasing its value in the form of a profit. There is no moral incentive to (freely) share data, even if it makes the job easier. Onerous Information & Communications Technology (ICT) contracts can place burdens on parties in how data is exchanged, and who holds ownership, copyright and intellectual property. Now I am not condemning them, of course not, they have heralded and made the availability of sharing data possible.

The development of smart buildings also has effects on automated systems, intelligent building management, adaptive energy systems, assistive technologies, remote monitoring and the Internet of Things (IoT) allowing feedback to be collected and the data harvested. Considering that the design life of a building may be 50 years or more, ICT in its current form, might become redundant before completion and occupation.

This becomes a bigger problem on an urban scale, where major infrastructure programmes will impact on the economy for years, but an ICT product may last only as long as procurement before it becomes obsolete (Designing Buildings Wiki 2019). Insurance-backed alliancing addresses this conflict but remains largely passive in an incentive or reward scenario.

There are many advantages with the implementing smart technologies in this building, The Edge (Cala-or 2023):

- Healthy indoor environment, with improved air quality and increased access to natural daylight which makes it a better indoor environment.
- Improved user satisfaction is shown where buildings adapt to users' needs so that comfort increases the user's satisfaction and can lead to higher motivation and productivity.
- Enhanced convenience, with the use of smart technology, it is easier to locate meeting rooms and improve collaboration among colleagues.
- Flexible and efficient space management, smart systems help utilise space effectively for various purposes.
- Cost reduction, sensors help the building structure reduce the cost of energy and heating and can contribute to long-term sustainability.
- Increases the property's value, intelligent buildings attract more tenants and buyers and can lead to higher unique selling propositions (USPs) and more profits for developers (Brahney 2019).
- Provides new business models which encourage designers and real estate developers to employ technology to improve operation efficiencies and user experiences.
- The management of the building has a clearer understanding, and real-time data provides a better understanding of building activity to the FM team for early detection of building issues.

There are challenges too where first and foremost is interoperability. The deployment of IoT remains challenging for PropTech (Property Technology firms) since it requires a seamless exchange of data and communication across different devices from various manufacturers. As more devices from different manufacturers are used, the complexity increases. The lack of interoperability can result in a waste-of-time and resources. Thus, this requires standard protocols like Building Automation and Control Network (BACnet) that coordinate with technical standards.

Compatibility with incompatible technologies, and complex building management system (BMS) networks can often hinder IoT applications. To overcome these challenges, it is crucial to integrate IoT connectivity solutions, this is the most reliable and cost-effective solution required by PropTech. Complex data management systems offer a vast amount of data generated from IoT which requires a lot of work regarding where to store the data, and who is responsible for it.

On the other hand, this can be addressed with a common platform that all actors can access and data analytics.

The integration of different systems should be scalable to cope with changes in the buildings. Many investors and developers cannot see the value of intelligent building technology, sometimes they view it as an unnecessary expense or a complex technology with drawbacks. Consultants and innovators often face challenges in convincing stakeholders of the benefits of smart building. While it can be costly to invest in IoT solutions, the economic rewards of making buildings more efficient are worth it in the long run. According to real estate is rapidly changing due to the higher demand for high-quality buildings, driven by EU regulations promoting energy neutrality combined with post-pandemic, they are starting to realise investing in smart buildings.

Promoting Motivation

It is worth repeating that pedagogically The Copenhagen School of Design and Technology course is structured through group work in a matrix diagram on project driven semesters. It is practice orientated and uses problem-solving methods. This is a huge benefit where collaborative work is involved and this is the case with BIM. The work can be divided into two parts, one where authorship is to the fore and the other where analysis is primo. Authoring involves building the model and developing it through the various work stages of the project. Analysis allows the model to be checked and controlled so that certainty is achieved, bringing projects on time and to budget. Allowing the data generated to be mined and tested is not new, it is in fact an integral part of the planning process and of great concern to the client.

The school's course syllabus is making inroads in this direction, looking to introduce an appreciation of these mechanisms in the first semester to full management of them in the fourth semester. Having a matrix organisation within the classroom allows this to happen so that the differing roles can be played out and their integration fully experienced.

Two things have emerged from this arrangement, first (and the one that was not anticipated) is the reduction of stress levels coming up to evaluations and exams. Each group can see the model, address its problems and contribute to the group work in a meaningful manner. Second communication is incredibly better, both internally and externally. The model is usually in place much sooner in the semester than previously and this allows the more difficult operations more time to be completed. Previously, right up to exams there was panic and grey areas where the design had not been fully resolved. While it is not a cure all package, more students can address the problems and assess their impact easier than before.

We live in a changing world and our students are going into the marketplace not knowing many of the jobs and careers that they will have to master. Some of which have not even been defined yet. Preparing students for this scenario means giving them the knowledge, skills and competences so that they can interpret and perform in these changing circumstances. I have always sought to get students to think for themselves and when confronted with problems not to have a rigid foregone solution, but to know how to go about finding a better-informed solution. This ability to adapt and innovate will, I hope, serve them well.

Latent Redundancy

A student of mine wrote recently:

> '. . . *Additionally, individual barriers, such as a lack of knowledge, skills, and resources, can further hinder the effective implementation of sustainable values in the (AEC) workplace*' (Brown 2023).

She identifies this through a lack of awareness, short-term focus, lack of incentives, pre-approved solutions, outdated industry values (and ideologies) and ultimately cost concerns. Similarly, given that projects are driven by contracts, and these tend to involve different interests, goals and values, there are opportunities for conflict. This can end in human error or even corruption and collusion (Montague 2023).

> '*For the last two decades, the (construction) sector has tried to apply the collaborative mantra. But at the end of the day, when the chips are down, it is the contract that shapes behaviours and outcomes*' (Kinnaird and Geipel 2017).

The solution craves openness, transparency, honesty and immutability, Arup's tells us. Blockchain offers permanent, secure and valuable transaction methodologies, they add. Don Tapscott sees the difference as stark as that of the internet of information, compared to the internet of value (Tapscott 2016b). Arup's goes on to describe a blockchain of circular BIM things. This is a *live* building information model,

> '*Whose components could continuously be fed usage data from real building throughout their operation*' (Kinnaird, Geipel 2017).

Using Levels of Development (LOD's), this is akin to going from LOD-300 to LOD-400 to LOD-500 and/or even LOD-350 in between (BIM-FORUM 2017). LOD-300 is the level of development that broadly defines the designers' requirements. 350 is the contractors' take on the designer's requirements. 400, then, is what is ordered to be built, while 500 is the *'as-built'* completed project.

Arguably, there should be no changes through these stages, but this is clearly not the case, as can be seen most glaringly in a project like Grenfell Towers in London (Moore-Bick 2019), where the project went through too many iterations from the architect's intentions to the completed refurbishment, resulting in disaster. New replacement windows were installed 150 mm from the position of the original windows. These gaps were not properly filled out, being filled with polyisocyanurate (PIR), which is combustible and clad in an aluminium composite sandwich panel filled with polyethylene (PE) cores allowing the fire unabated access to all areas (Barratt 2018). Polyethylene is described as a solid petrol substance in a fire (Booth 2022).

Remedial Action

Having a ledger to document this process, and having it decentralised would make it transparent. But it does not end there, by having methods to monitor the use of the building and sensors to report this data back to a database means that algorithms can process this information and establish if the building is delivering what was claimed at the design phase, i.e. accountability is introduced to the contractual obligations, which was not there before.

Being able to measure these savings (should they accrue) in energy use, embedded carbon and optimal occupancy, allows for them to be documented and a method of rewarding such endeavours becomes real. A corollary of this is that once this is realised, then the actors in the design team and construction team, will find ways to improve this new vertical source of income, becoming better at buildability, sustainability with better performing building stock. Thus, they become incentivised (Mathews et al. 2017).

David Cumming writing an article on why asset managers cannot be passive on climate change said:

> *'The obvious template for solutions is the 2015 Paris Agreement, which seeks to limit the increase in global temperature this century to well below two degrees Celsius above pre-industrial levels, and to pursue efforts to limit the increase even further to 1.5 degrees Celsius. Investors*

should recognise we are nowhere near these levels currently: the FTSE 100, for example, is on course to burn the planet at 3.9 degrees Celsius' (Cumming 2020).

He seeks a solution that is responsive that is substantive, authentic, informed and impactful. To do this means putting pressure on companies and governments to enact policies to deliver these objectives and concludes:

> *'Climate change has changed everything. Investment objectives now have to include responsible values and actions, in addition to financial returns. We have to respond by engaging in a different way and by taking decisive action when the companies we invest in don't. We cannot be passive in the face of climate change. We have to be active'.*

Daisy Dunne, writing in The Independent (Dunne 2020), notes that the ten most destructive extreme weather events in 2020 alone cost the world about $140bn. These included the bush fires in Australia, extreme flooding in Asia, and severe storms in North America. Because of unusual rainfall in the Middle East and Eastern Africa, plagues of locusts did untold destruction across Africa. In 2018, KPNG reported that natural catastrophe-related economic losses reached $160 billion. The vast majority, 95%, of the registered events were weather related (Brown 2019).

Finally, In the World Economic Forum report of 2019, extreme weather events and the failure of climate change mitigation and adaptation, are seen as the most likely to happen, together with having the biggest impact of all global risks in their landscaping figure (Beswick and Bailey 2019). If there is any good news in all this doom, it is that insurers are telling, no screaming, at politicians that this cannot go on, and that measures need to be applied, otherwise they cannot offer adequate cover. Without insurance, the wheels come off.

Insurance Premiums

If any industry or sector can change the ultra-right-wing conservatives into adopting climate change as real and costly, then it is insurance. Ernst Rauch, Munich Re's chief climatologist, has warned that climate change could make cover for ordinary people unaffordable after the world's largest reinsurance firm blamed global warming for $24bn of losses in the Californian wildfires (Neslen 2019). Furthermore, he pointed towards wildfires, flooding, storms and hail meant that the only sustainable option was to increase premiums accordingly. There is a consequence too for people on low or average wages not being able to buy insurance. He continued:

> *'The lion's share of California's 20 worst forest blazes since the 1930s have occurred this millennium, in years characterised by abnormally high summer temperatures and 'exceptional dryness' between May and October, according to a new analysis by Munich Re.'*

Many hold that this will be the prime method, forcing unconcerned governments towards more sustainable options. Already repeat premiums are rising in areas prone to flooding, although I know of no-one being refused cover to date. Passau a city 200 km from München on the Danube is a good case in point. In 2013, the river rose to the highest level ever, peaking at 12.8 m, beating the previous best of 12.2 m in 1954 and the worst in over 500 years (Reguly 2013). Reguly wrote in that article that:

> *'Munich Re, the German giant of reinsurance—the business of insuring the policies of insurers - put the damage from the June floods at €12 billion. It was estimated that insurance covered only about €3 billion of that amount, meaning a lot of people were out a lot of money. Munich Re's primary insurance arm, ERGO, alone paid €83 million in claims to German flood victims - twice as much as for the last big flood, which was in 2002.'*

Peter Höppe, head of geo-risks research at Munich Re says *'I'm quite convinced that most climate change is caused by human activity'*. Such a statement is remarkable, given that currently American insurers will not mention *climate change* and *anthropogenic* in the same sentence, according to Reguly. He also points out that Munich Re first became aware of this a far back as 1973 when increases began to rise.

> *'How did Munich Re and the other reinsurers get it right so early? The answer, in a word, is fear - fear of losses that could destroy their business. No industry has more incentive to know the effects of climate change than the reinsurance and insurance industries'.*

Worldwide, the insurers pay weather-related claims of about $50 billion (U.S.) a year, a figure that has more than doubled every decade since the 1980s, adjusted for inflation. In 2012, the 10 costliest natural catastrophes, from Hurricane Sandy to floods in Pakistan, caused $131 billion (U.S.) in damages, of which about $56 billion (U.S.) was insured, Munich Re says (Reguly 2013).

CHAPTER 9

Blockchain

Construction is generally perceived to be fragmented, largely because it is bedevilled with notorious low (profit) margins, due the traditional tendering process, meaning the lowest bid usually wins the project. Therefore, it often pays twice to correct poor or faulty workmanship (Harty 2012). John Egan (1998) and Michael Latham (1994) correctly cited these evils within construction and identified corrective measures, but without incentive or reward (Mathews 2017; Mathews et al. 2017). Arup's also acknowledged the same problem in their report about Blockchain Technology:

> '. . . *However, the sector is limited by the existing data processing and exchange methods which remain characterised by analogue methods that support old adversarial behaviours*' (Kinnaird and Geipel 2017).

Whenever a project is handed over, there is a procurement process that now reaches the end of a cycle, mentality. Increasingly, there are methods being tried that seek to prolong or engage that contractual agreement, whether it is soft-landings, life-cycle assessment or

Transforming the Construction Industry with Blockchain: Enhancing Efficiency, Transparency, and Collaboration, First Edition. James Harty.

facilities management. What none of these techniques embrace is unpredictable performance. There is also no incentive to do so, or a reward system in place to encourage better performance.

The adoption of blockchain-enabled services will be a major player in the transition to automated services (Fitch 2023). It is also a mistake to consider blockchain as a single monolithic technology. It provides ways for diverse parties to work together, while coordinating their systems and streamlining the methods. It will also facilitate digital transformations in procurement. It will prompt us to rethink many procedures in how we deliver services It is more reliable at record keeping, increasing security, allowing more transactions.

Blockchain-enabled processes have the ability to embed and automatically execute smart contracts. So, when a set of predetermined conditions are met it runs. This excludes manual triggering, or a slow timeline, meaning better efficiencies and a smoother ride for all involved. For someone providing a service or product, this is heavenly. For the receiver, this removes the ability to delay payment, which might be perceived as a disadvantage or be contrary to long held practices, or corrupt tendencies, like reducing payment or forcing extra work quid-pro-quo.

We have seen unscrupulous clients deeming exposure or involvement in prestigious projects as payment enough, meaning that in future work can bring prequalification or indicate experience in this particular branch. Preventing such situations means that money must be held in trust by the blockchain, meaning that it must exist before commencement of a project. This can play havoc with concerns which generate funds on the side or can only cash-in upon handover. Borrowing, brings financial institutions into the picture, and this might discourage collaboration.

A compendium of principles as an application allows the stakeholders to value themselves, but value is only added to projects through people (Smyth and Pryke 2008). Therefore, the management of relationships becomes very important. The construction industry is accepted as being fragmented, rarely do the same people work together on subsequent jobs and often they do not complete the current job through either disruptions in the work phases themselves or the sheer length of the project, which sees them either replaced of decanted to other projects.

Even a team kept together for more than one project will often meet new players as their opposite numbers subsequently. Even if the same companies remain involved, the personnel often change. In addition, the magnitude of small firms involved and the whole culture of sub-contracting out, engenders a state of flux, and conversely have a vested interest in protecting niches and (their perceived) expertise in the market. The corollary to this can be seen where key team members or personnel are explicitly written into contracts so that they are obligatory involved in the project. This comprehensively arises from clients' wishes alone.

But positive relationships do add value, improving project performance and client satisfaction. They also, as mentioned, induce less adversarial behaviour from the top-down and offer procurement-led measures for proactive-behavioural management throughout the enterprise. Relationship contracting is currently best seen in partnering and supply chain management.

One of the best examples is Terminal 5 at Heathrow (T5), completed on time and to budget, opened in 2008, which is rare for a building of it size and complexity (Ferroussat 2007). T5 nurtured and encouraged such an environment (Latham 1994). It was based on the principles specified in the Constructing the Team and Rethinking Construction (Egan 1998). T5 could have ended up opening 2 years late, with 40% cost over runs and up to six fatalities (Potts 2009).

Conversely, Eurotunnel had difficulties in motivating the suppliers once the contract had been awarded. Winch calls this 'moral hazard' (Winch 2002), where the client is somewhat unsure that the contractor will fully mobilise its capabilities on the client's behalf, rather than its own interests or some other third party. The preferred option he calls 'consummate performance' instead of more likely 'perfunctory performance'.

The root to this situation can be found in the negotiation of the contract, essentially between banks and contractors. Here two cultures collide, on the one hand; the banks prefer to move the contractor to a fixed price, to reduce their risk. On the other, the contractor works on the basis that the estimates have to be low, to ensure that the project gets commissioned.

> *'In banking you bid high and then trim your margin: in contracting you bid low and then get your profits on the variations' or as another said 'the project price . . . was put together to convince the governments, it was a variable price, a promoter's price. What it was not was a contract price'* (Winch 2002).

The economic, information and material flows in construction logistics are usually disintegrated, writes Dimosthenis Kifokeris and Christian Kock (Dounas and Lombardi 2022). Blockchain, it is suggested, could integrate them all into a shared digital ledger, using smart contracts to create value to all stakeholders. This facilitates transactions, flow integration, interactions and trust development. They can be event-driven or event-inducing, which raises an interesting position of interdependence and proactive initiative.

This last point means that the enterprise begins to play with the data, analysing patterns and formatting the data into new worlds. Being driven is a new aspect which adds value. It will inform better practices and plot new courses of action. In the bigger picture space, this tells the user if they are doing it right, can they improve what they are doing or is a new course now needed. By inducing, it is meant that the application might be plotting for itself, offering new thoughts to the user. This has not been available to us before. It is a new paradigm and will transform the industry.

Provenance

Blockchain can track the provenance of data used to train machine learning, which means that it is reliable and trusted. This means that in a new era where the data cannot be verified (as in fake news), blockchain ensures that the data has not been tampered with. Making the data reliable makes it not biased too. It can also track and audit the data throughout its lifecycle and report on its performance so that any problems can be easily identified and corrected before much damage is caused.

Provenance is evidence (and a complete documentary) of an entity that can be used to prove its legitimacy, if required. It captures the route mapped out by its history (including ownership). This has a major impact on supply chains, meaning that the sequence through which something has travelled can be documented and traced if there is an issue in that chain. Because, there are many actors in construction, the likelihood and possibility of something going wrong increases exponentially.

A typical example of this is found in a supermarket where a batch of food might be found to be contaminated. Previously, all this product across the chain would have to be withheld until the fault was found. Now, there is a track record of this particular batch, and each step of its journey can be traced back to where the fault can be identified. The impact of this is immense, saving time, money and inconvenience.

It provides authenticity and trustworthiness of an asset giving it integrity and immutability.

This affects supply chain logistics, asset management, legal ownership and finally legal claims. It can trace multi-party transactions. It provides a footprint for smart contracts, and automatically timestamps and records all transactions in a distributed ledger. Moreover, it is secure and has configurable access rights. Coupled up with artificial intelligence this process can be fully automated, making it robust and resilient in the field.

Cryptocurrencies

Cryptocurrencies have three features: ensuring pseudo-anonymity, independence from a central authority and double-spending attack protection. Several coins have established themselves since the emergence of Bitcoin on the 3 January 2009, most notably Ethereum, Litecoin or Ripple. Satoshi Nakamoto is credited with developing Bitcoin and devising the first blockchain database. It is a presumed pseudonym that was allegedly born 5 April 1975 in Japan (Wikipedia 2021).

A Bitcoin is a virtual cryptocurrency, and its address will consist of a random code of between 27 and 34 characters, for example:

13h1ULGwrKYTelLwged98oTzjG4SShQjNUh (Harkell Short 2014).

The latest Bitcoins are logged on a web page called www.blockchain.com/explorer and can be seen as they are mined every 20 minutes or so. Their value accrues as they are transacted for Fiat currencies (Latin for 'Let it be done' as in an order or decree). Money per se is a commodity that is underpinned by some physical good such as gold. Fiat currency is underpinned by the regime that issued it such as the Euro (€), US dollar ($), Pound Sterling (£), and many others. This means quantitative easing (QE) can be exercised. QE is a monetary policy where a central bank buys long-term securities on the open market in order to increase money supply to encourage lending and investment. This stimulation of the economy has been criticised as benefitting the rich, increasing income inequality.

Cryptocurrencies have three features: ensuring pseudo-anonymity, independence from a central authority and a double spending attack protection. Several coins have established themselves since the emergence of the Bitcoin on the 3 January 2009, most notably Ethereum, Litecoin or Ripple. Satoshi Nakamoto is credited with developing Bitcoin and devising the first blockchain database. It is a presumed pseudonym, who was allegedly born 5 April 1975 in Japan (Wikipedia 2021).

When Bitcoin first entered our daily lives, it had no value. It was put into circulation as a *reward* for using the computer's computing power for solving complicated mathematical problems. This is important, as it is the number crunching aspect of the process that is generating the value, but as said, at the start, the value is zero. As a digital asset, it worked as a medium of exchange. By definition, it meets Jan Lansky's six criteria to be a cryptocurrency, which are:

1. The system does not require a central authority; its state is maintained through distributed consensus.
2. The system keeps an overview of cryptocurrency units and their ownership.
3. The system defines whether new cryptocurrency units can be created. If new cryptocurrency units can be created, the system defines the circumstances of their origin and how to determine the ownership of these new units.
4. Ownership of cryptocurrency units can be proved exclusively cryptographically.
5. The system allows transactions to be performed in which ownership of the cryptocurrency units is changed. A transaction statement can only be issued by an entity proving the current ownership of these units.
6. If two different instructions for changing the ownership of the same cryptocurrency units are simultaneously entered, the system performs at most one of them (Lansky 2018).

Thus, an intrinsic value has been found, for a previously intangible value, and now it is performance-based, which can be rewarded.

AECcoin

Once collaboration is fully implemented, methods will be needed to codify good collaboration and discourage poor performance. One such method could be a distributed coin awarded for good behaviour. Often in collaborative environments, methods are needed to promote and procure better practices. Peer-to-peer pressure is often sufficient to encourage good collaboration but can be difficult to map out and adequately grade.

In the cold light of day, a grade cannot be given on how well collaborators have worked together. Of course, the proof is in the eating, but often these attributes cannot be bulleted and checked off as being met. Many methods have been proposed but few have borne results. One method is to get each group member to rate their colleagues, friendship often interferes here as most, out of loyalty, give each other top marks. If one member has not carried their weight, or has been onerous to the others, then that is easily exposed, but when a graded response is needed then those asking are not being heard by those asked.

Another iteration of this practice is the weighting of the answers so that differing attributes get differing rankings. The object here is that those features that are held highest should be rewarded best, but the whole process is very subjective, with little or no chance of being objective, no matter how much learning objectives have been mapped or charted out on a vanilla envelop.

Instead, having something of value to give to someone else as a reward for good collaboration is a novel idea and in a one-off situation does not bear much scrutiny. But, if the practice was spread over several projects, those performing best would accrue more of the said vouchers. The upshot of this would be that at a job interview, the candidate holding 20 of these would more than likely get the job over someone holding three.

Bitcoin earns its stripes by the mining process which makes it rare and definite. Here the earning potential is the hard-work shelled out in the collaborative environment. This means it has an inherent value, and an intrinsic value, making it valuable. Malachy Mathews has coined the term AEC coin to chart this process (Mathews 2017). The proof in the pudding is how seriously it is taken. Bitcoin had its teething problems and was abused initially, with the animosity of the coin making it of interest to underworld and drug related groups.

But it came through this and is thriving today. Precisely, because it is robust, transparent, and resistant to modification. Being an open distributed ledger gives it the foundation to make it reputable which makes it acceptable and trustworthy. The same applied to AEC coin would imbue it with the same properties and give it a credibility. An incubator space could be architectural technology courses where the students, in groups, were awarded a coin at the end of the semester. Initially, this might be a slow burner, but by the time internships happen, or gainful employment, their worth would blossom.

Some schools hold a graduation show, where graduates are given a wall space to promote their project. Usually, a small green baize panel is placed at the bottom right of the presentation, for third parties to leave their business card. The cumulative effect of this, is that the panel that receives most cards, also garners passing interest, the thought being this must be interesting. This then informs the graduate of their worth or value and encourages them to follow-up on the initiative. Both the show and the panel are very much analogue. An AEC coin would digitalise this process.

CERTcoin

A cryptocurrency, that recognised and rewarded learned work, with a learning record statement, mapped a learning process and offered secure access and a robust ledger, would have great merit. Having a platform that provided learning modules, which could be offered and once completed, award badges or recognition certification, which could be earned to show achievement, is something most students readily buy into and something that is increasingly seen as a new way of learning is gaining traction.

Many such platforms exist, whether it is BIMcert, Lynda or Microsoft Learning (to name a few), and their most beneficial attributes are that they provide just-in-time (JIT) learning, in short manageable bites that appeal to the perceived audience today. Their downside is that they cannot, or are poor, at building loyalties or repeat business. In short, they are used when there is a problem, but they are not the go-to app which they would like to be, for each and every request.

One method of correcting this shortfall is the introduction of a dashboard and a gamification element to encourage peer-to-peer competition, and this is to be lauded. But being number one, while challenging in itself, cannot be taken out of the system, or easily be recognised by third party employers or would-be clients. A method is needed to reimburse the work carried out, and it needs to universally respected and transcend national boundaries, especially in Europe, where freedom of movement (employment-wise) is guaranteed.

Blockchain offers three elements to promote these requirements. First, blockchain has a trace and trace ability, which is a real-time method of showing where the student is on their learning path. Second, it can be a ledger, noting what a person has learned, without be compromised. Lastly, it can reward such practices with a coin that the student earns for completing modules, which can be guaranteed evidence to would-be third parties. While sounding relatively nominal and simple, this ability is intrinsic in a method needing transparency demonstrating incorruptibility and a robustness that stands up to scrutiny.

Digitalisation, within the construction sector, is increasing at a welcome pace, and not before time. One of the most underestimated aspects of this surge is that it is ripe for what is known as BIM Level Three (Bew and Underwood 2010). Level zero refers to pre-modelling times, where drawings and documents might be digital, but in name only, data had no underlying value, by being shared or reused. Level One saw the transition from 2D to 3D, but still no value was attached to the models developed, beyond delivering contract drawings and documentation. BIM Level Two became the minimum requirement to meet digital formats in 2016 in the UK with the introduction of the digital mandate, where all public tenders had to be digital. Initially, this could be an excel spreadsheet, but essentially, the assets could be shared and inspected.

It caused a rush to set up protocols and standards, most notably, BS-1192 series. Publically Available Specifications (PAS) 1192-2 described standards, specifications, codes of practice and guidelines to implement this new paradigm. They were critical in writing and producing Information & Communication Technology (ICT) agreements and embraced Building Execution Plans (BEPs). They were to become an International Standard (ISO 19650).

BIM Level Three engages in the data accumulated and begins a process where it is analysed, performance can be checked and simulations can be performed, leading ultimately to a digital twin where a real building and its virtual twin can inform and predict how the building is performing and how it might be tweaked to perform better or under other circumstances. This is a proactive course of action and moves us on from passive monitoring of a buildings impact. This means that the building's life can be monitored, and sensors can give feedback about the use of the facility while beginning a process of having a smart facility in a smart city environment.

This Tsunami of data needs filtration to make sense of it so that it presents cases on which better informed design decisions can be made. The Internet of Things (IoT) has begun the monitoring of performance whereby the building can inform how it is being used. Artificial intelligence (AI) can also make informed decisions on our behalf, and using algorithms, predict and maintain best practices. This forth industrial revolution, as it has been called, is rolling out a whole new plethora of methods and mechanisms, which need controlling and understanding, in order to keep everything on track. Blockchain has an important role to play here, the management of all these new features, requires careful supervising and qualified handling.

Through blockchain technologies, value creators, such as designers and learners, can display or transfer value to their clients and employers directly. This is an intrinsic value. These values are brought to the table because of blockchain. Without it, no value, with it the improvement to the service brings certainty to the proposition, meaning all concerned can operate with confidence, knowing that risk is significantly reduced.

This can also be seen when tasks are completed, blockchain can host an interface that vouches for the work, releasing payment or reward, as appropriate. It can all happen seamlessly, independently to other stakeholders. This removes delays in payment, ensures that deadlines are met and rewards the efficient management of workloads. It can track and trace intellectual properties, protecting both producers and consumers of products and services. This restores who and what others can see by allowing the owner of the content to decide how that content is used and/or abused.

Through the use of blockchain platforms, users can both use the service and enjoy additional benefits by participating in the management and control of the network. Additional benefits include verification of what you have earned to control as to how it is managed and to whom it is available. While sounding irrelevant, it is crucial to how your data is shared or displayed.

Crypto Currencies, Historical Relevance to Form a Foundation of the Work Required

As far back as the early 80s, the internet had major problems with privacy, security and inclusion. Often because of third parties, too much personal data was disclosed, and scalability raised issues with micro-payments. Fast forwarding to the crash in 2008, and Satoshi Nakamoto outlined a peer-to-peer protocol for electronic cash which became known as Bitcoin.

This is a cryptocurrency and what makes it different from other currencies is that it is not controlled by banks or sovereign countries, meaning that it rides above untrustworthiness. This is critical in understanding the impact of such a statement. Back in the fourteenth century, a double-entry accounting system emerged from merchants and money lenders in northern Italy.

It involved a ledger where each transaction was entered twice, once as a credit and once as a debit, with all entries leading to equity and a bottom line of zero. Luca Paciolo invented double-entry accounting allowing enterprises scalability. It brought accountability to accounting and this is a good thing. Where things go wrong is how it is managed, typically that the books are in order and not cooked. This can be a trust problem or human error. It can also be fraud or that the scale of the transactions become untenable as a micro-transaction.

Because it is a third party, it could also be open to manipulation. This might mean that management has not acted with total integrity or has become a tad greedy. Cronyism, corruption and false reporting can cause items to be inflated or deflated, to be assets or pure risks. These are all value statements and can become very attractive inviting prying eyes.

Human error is the leading cause of accounting mistakes. Nearly, 28% of professionals reported that people plugged incorrect data into their firm's enterprise system. If anything, the growing complexity of companies, more multifaceted transactions and the speed of modern commerce create new ways of hiding wrongdoing. Traditional accounting methods cannot reconcile new business models, micro-transactions for example.

Blockchain can document transactions, which are then chain linked so that they cannot be tampered with retrospectively. The resulting chain of blocks is held by the stakeholders, making it decentralised and most importantly robust. It removes the third party whether it is the bank, an agency or a vendor who is required to validate a transaction. They typically required specific information, charged transaction fees and kept information about you. Bitcoin sough to eliminate the middleman (Forester 2018).

Sidechains allow value-compatible assets to migrate bi-directionally. This allows new or alternative ideas to be rapidly explored and be brought into the chain or discarded, allowing evolvement to meet stakeholders' demands. A claim towards sustainability can now be measured over time and be paid out or rewarded as agreed or plain performance measured. Lies or untruths are exposed and dealt with accordingly.

But this regime requires management (middlemen), leading in some cases to cronyism, corruption and false reporting precipitating bankruptcies, job losses and market crashes (Tapscott 2016b). If the management is automated, then it is more:

- Resilient
- Robust
- Real time
- Responsive
- Radically open

- Renewable
- Reductive
- Revenue-generating
- Reliable

What is not to like?

Data is a commodity, it has value. Using BIM based collaborative technology on a traditional cooperative process, enforced by 'self-serving' contracts, is doomed to failure. Interaction with the BIM database is an activity that generates economic value, which may be measured into existence by an electronic token that rewards disassociated parties for maintaining and improving the database for the benefit of all, thereby replacing the third party intermediary characteristic of legacy hierarchies with a simple and efficient 'digital handshake'.

Hierarchical management structures are being replaced by network structures in many industries simply because networks are more efficient, enjoy higher market valuation, they are fault tolerant, and self-regulating, whereas hierarchy requires substantial managerial and administration overhead to secure individual nodes.

The construction industry is highly fragmented and has been deplored for being very adversarial. Construction owners are risk evasive, while contracting parties interpret contract clauses differently and for their own benefit. Productivity levels are low compared to other industries and have even dropped over time in some countries (O'Connor 2009).

The notion of BIM was that 3D models would be an efficient way to produce 2D documents, the next evolution of CAD enhancement. But it quickly morphed to a point where the model created brand new value networks: clash detection, quantity take-offs, field BIM, direct fabrication, energy analysis and, ultimately, BIM models as a store of a myriad of facilities management information mapping the lifecycle analysis. AEC professionals have always struggled to recover the intrinsic value of their labour. Repeatedly, they work on small margins, with no recourse to added value.

Blockchain with its properties of transparency, immutability and consensus validation now offers AEC professionals an opportunity to develop a 'new value proposition' to extract reward not only just for their collaborative services that they have provided but also the intrinsic intangible value of their collaborative professional service over the lifecycle of a building.

This proposes a 'new value proposition' for clients and AEC professionals who see the benefit of working together now have at their disposal an array of collaborative tools to allow this to happen in real time. Blockchain is a robust technology that will record value transactions of the collaborative effort and provide a method to reward intrinsic value through cryptocurrency. It becomes the gift that keeps giving. It becomes the contract that rewards value. It is also a contract that does not reward non-performance, and an agreement that releases payment when the contract is met.

A combination of BIM and blockchain has the potential as a platform for true collaboration where visual evidence of 'value transactions' is written into a ledger, timestamped, gathered and through consensus locked into a block, visible for the stakeholders to see. A platform like this will disrupt the design and construction industry.

D'APPS, Most Interesting in How Blockchain Will Infiltrate Much of Our Everyday, Which Is Not Appreciated Today as It Should Be

Today, apps are a part of our everyday, having infiltrated our telephones (mobiles), IoT and many other aspects of life. A mobile application is a computer programme or software application designed to run on devices. Initially, they were thought to be helping hands to better productivity in accessing database. In 2010, it was listed as the word of the year in the American Dialect Society.

They can be categorised into native apps, hybrid and web apps. Native apps are specifically for handheld devices, web apps run on a browser, while hybrid, as the name suggests, run on both. A typical pre-bundled app is Google Maps on a mobile phone. It can show you where you are and direct you to a new destination, finally, it can offer several modes of travel with estimated times of arrive, cost and also report on the ease of use, with regard to public transport's capacity or a combination of suggesting the bus some of the way and walking the remainder.

Most apps are free, but they might harvest your data as a price, and as a vertical growth object they can make people very rich. They are backed by algorisms that connect data, discover patterns and promote particular ideas or doctrines. Understanding their power is important, and often their potential was only realised after their release, when their impact raises the bar to new levels. With these new powers come exploitation and whistleblowing exposing the depth of their reach.

Cambridge Analytica is a prime example. In his book, 'Mindfu*k, Cambridge Analytica And The Plot To Break America', Christopher Wylie reveals the data-mining operation that Cambridge Analytica engaged in to affect the 2016 American Presidential and the 2017 Brexit Campaign, by harvesting 87 million Facebook accounts globally (Wylie 2019). Recruited by Alexander Nix, then at Strategic Communication Laboratories Group (SCL) in 2013, their scope of work included behavioural research and strategic communication consultancy.

In her book 'Targeted My Inside Story of CAMBRIDGE ANALYTICA and How TRUMP, BREXIT and FACEBOOK Broke Democracy', Brittany Kaiser, covering the same topic, wrote 'In the digital age, data was the new oil' and 'that data was an incredible natural resource. It was the new oil (again) available in vast quantities, and that Cambridge Analytica was on track to become the largest and most influential data and analytics firm in the word' (Kaiser 2019).

She writes that Facebook shared information using Friends API's, 'a now notorious data-portal that contravened data laws everywhere, as under no legislative framework in the United States or elsewhere is it legal for anyone to consent on behalf of other able minded adults' (page 79). On 28 April 2015 Facebook shut it down, but the damage had been done, and 240 million Americans (out of 328) had had their data harvested, amassing over 5,000 data points per person. This allowed them to profile people using behavioural microtargeting, which ultimately allowed them to get certain swing voters to vote for their candidate, or worse still discourage others from going to the polls at all.

Cambridge Analytica's chosen method of operation involved having clearance to secret level information, whilst their members included former cabinet ministers, retired military commanders, including professors and foreign politicians. They worked for military organisations and intelligence agencies, conducting psychological escapades while their scope of influence knew no boundaries.

Later that year, Wylie was sent to Cambridge to meet with Steve Bannon, a political strategist for Donald Trump. They talked about research on cultural change. Wylie went on to show a map of Trinidad, where a layer of dots populated the map. Wylie adds: 'Those are real people . . . we have demographic data on . . . (their) gender, age and ethnicity'. A second click revealed their internet browsing, followed by census information and social media profiles. Bannon asked could this be done in America.

The relevance of this is not lost on the climate crisis. Donald Trump on becoming President of the United States in 2016 has systematically removed many critical measures designed to halt or limit environmental issues. Ninety-five environmental rules have been rolled back under Trump (Popovich et al. 2019). Twenty-five in air pollution and emissions, 19 in drilling and extraction, 12 in infrastructure and planning, 10 in animal welfare, 8 in toxic substances and safety, 10 in water pollution and 11 others.

Many of Trump's appointees 'are buried in layers of shell companies, creating vast opportunities for conflict' with minimal transparency (Confessore 2017). He is 'populating the federal government with former lobbyists who in many cases are influencing policy in industries where they were recently paid' (Lipton et al. 2017).

The National Climate Assessment, America's report on climate knowledge, was also prone to such attention. Hundreds of scientists in government and academia compile the best insights available on climate change, and this is released twice a decade to inform the government decision making process. In 2018, officials tied to lean on top scientists to soften their advice, without success as they refused. They later tried to bury the report releasing it after Thanksgiving, when there is little media attention, but that did not work either.

> *'Thank God they didn't know how to run a government, . . . It could have been a lot worse'.*

So said Thomas Armstrong, who during the Obama administration led the U.S. Global Change Research Programme, which produces the assessment (Flavelle 2021).

Dr Stuart Levenbach, a senior advisor to Trump's White House tried to tone the document down, whereas Dr Virginia Burkett, chairwoman of the Global Change Research Programme, was forced out. Trump, when asked about the findings that could devastate the economy, responded, 'I don't believe it'. Sarah Huckabee Sanders, said the assessment was 'not based on facts', while, Ryan Zinke, who was secretary of the interior, said that its findings emphasised 'the worst scenarios'.

Dr Michael Kuperberg, a climate scientist from the Department of Energy, was replaced by David Legates, a Trump appointee at NOAA who previously worked closely with groups that deny climate change. Another, Ryan Maue, who has criticised climate scientists for what he has called unnecessarily dire predictions, was moved to a role in the White House that gave him authority over the climate programme.

David Cameron is also embroiled in an embarrassing lobbying exposure, where he was seen to try and use his former position as UK

Prime Minister to influence Boris Johnson's Chancellor, Rishi Sunak, into giving lucrative contracts to Lex Greensill during the Covid 19 epidemic. In the run up to the 2017 general election Daniel Green donated £135,000 to gain access to major government departments (Devlin 2021). Greensill went bust in March 2021 was once worth $30 bn, employing 440. They lent money in the supply chain finance sector, was founded by Lex Greensill in 2011 (Harley-McKeown 2021). Cameron made about $10 m in salary and bonuses for two and a half year's work (Allegretti 2021).

With the outbreak of COVID-19, the UK government announced a financial aid package named the Covid Corporate Financing Facility (CCFF) of about £330 bn, of which Greensill sought to receive £10–20 bn of the package (ibid). It was refused but both Cameron and a specialist advisor Sheridan Westlake followed up with: 'What we most need is for Rishi to have a good look at this and ask officials to find a way of making it work. It seems nuts to exclude supply chain finance'. (Makortoff 2021). In the same article, Cameron also went to Saudi Arabia with Lex Greensill to meet crown prince Mohammed bin Salman, who was accused of approving the death of the journalist Jamal Khashoggi in the Istanbul embassy.

An investigative inquiry was called by Boris Johnson, to look at the problem of government contracts and lobbying. The chair was awarded to Nigel Boardman (of Slaughter and May, but in a private capacity). Boardman himself has received 10 contracts worth over £7m since March 2020 (Elgot 2021). The Guardian reported that the firm were 'working as an integrated team with Treasury legal advisers', as the Treasury set up the CCFF, which cost £400,000, according to the analysis of government contracts (ibid).

The shadow chancellor, Rachel Reeves, has claimed that the Conservatives are set on glossing over this cronyism, acting as if they can carry on like nothing has happened. Finally, Slaughter and May challenged the Cameron administration, when as prime minister he proposed changing lobbying rules to avert this very situation.

Michael Gove, minister of the cabinet office, has also being outed for misleading parliament over the award of a £22.6m contract to Bunzl Healthcare for personal protective equipment during the COVID-19 pandemic.

> *'Tory peer Lord Feldman lobbied for the contract last March while acting as a government adviser on PPE – whilst also being paid by Bunzl as a client of his lobbying firm'. Angela Rayner, Labour's deputy leader, told The Independent: 'Ministers have finally admitted what we all know*

to be true – if you're the mate or a client of an influential Tory then the rules don't apply, and you get to the front of the queue for taxpayers' money' (Forrest 2021).

We know what gets measured, gets done. Objectives and Key Results (OKR) syndrome. Now, the relevance of documenting this mess is that blockchain might offer a method of regulating these phenomena and bring calm to the cauldron. It can bring veritablity to the table while calling out fake news and fraudulence.

Smart Cities, the Culmination of How These New Technologies Will Impact Architecture, Construction and The Whole Building Sector

Looking at a city, such as Como in northern Italy, the 30-minute rule applies. This is the amount of time a person is happy to commute from home to work on foot. With the advent of the automobile, this 30-minutes commute expanded the scope of where home and work could be, and this expanded cities and urban areas. The intimacy of the old medieval city reflects on the eye-object contact that is necessary to establish communication.

Compared with how this translates or scales-up means that to catch the attention of the subject on a highway means resorting to in-your-face statements, best captured by Robert Venturi in 'Learning from Las Vegas' (Venturi et al. 1972), about being a monument or a decorated shed. This means great poles with huge 'M' for MacDonald's, to denote that the next exit, from the motorway, leads to a burger joint, or similarly huge billboards to convey messages at speed to passing motorists.

Today and we are increasingly seeing Uber and Lyft providing a service where it is possible to drive without having to own a car (Swisher 2019). Also, cars and electrical scooters litter city-landscapes where through an app. access can be gained and once the journey is complete the vehicle can be abandoned again. Sooner than you think, car ownership will disappear as these methods become cheaper and more accessible. Kara Swisher, writing for the New York Times says:

'Owning a car will soon be like owning a horse – a quaint hobby, an interesting rarity or a cool thing to take out for a spin on the weekend... But the concept of actually purchasing, maintaining, insuring and garaging an automobile in the next few decades ...'

The fact that she confines the horse to history, as a quaint thing for weekends, dressage, racing, and a trip to the country, consigns the once great beast of war to the rubbish heap. In the First Word War horses cascaded over the battlefields, only to be mown down by the ubiquitous machine gun. In the Second World War the French cavalry were useless. The horse's time was up. She sees this as being the same for petrolheads, the weekend only, removed from the day-to-day chaos of life.

She writes that car ownership actually declined last year and points out, in her defence, how quickly we have adopted map-apps over physical maps, snail mail to e-mail and watching on-demand to prime-time TV. The same happened with landlines and mobile phones. There is even a nod to declining retail shopping being replaced by online shopping and the quick delivery pioneered by Amazon. This too has a huge bearing on the high street and shopping malls, and by consequence how our cities will survive, as we know them.

Car-sharing services is also a small step towards a carbon-free life and with removing 80% of stationary cars off the streets, brings street-life back to centre stage. The street will prevail but the with a new lifestyle. More radical will be the movement away from four wheels to airborne drones. Already the technology is in place to avoid crashes with autobraking systems installed on front window screens. Human error is removed to a degree.

IoT is having an immense impact on smart buildings and by extension smart cities. Increasingly, sensoring is becoming inclusive and prominent in building management systems. These comprise of motion or occupancy sensors, indoor air quality sensors, room condition sensors and temperature sensors. Each has a role which collectively makes managing a building more comprehensive, both in its everyday use and long-term prognoses in how to best manage the project and how it can interact with its neighbours in a meaningful manner, making our communities better served.

Time of flight sensors note when a person enters or leaves a building, usually through an entrance mechanism via a card, dongle or by monitoring a hand-held device. Light and HVAC sensors control the lighting, heating and cooling systems, by turning them off when no occupancy is detected. Toilet flushes and hand dryer usage inform cleaning personnel where to direct their efforts. Automated doors detect a person's body temperature to open the door. Space optimisations track the presence of people in a commercial space.

Indoor air quality sensors monitor the quality of the air within a building. Often as the day extends air quality deteriorates, so it measures the amount of CO_2, volatile organic compounds (VOCs) and particular matter within spaces. Room condition sensors optimise room conditions measuring the light, sound, temperature and humidity of a space in real time. Finally, temperature sensors measure the ambient temperature of a room.

We spend 90% of our lives in buildings so maybe we should address their effect of us. It is important to feel comfortable and healthy in these fortresses. On the other hand, monitoring means we can measure the energy needs and allocate resources better. Air quality can account for 50% of sick leave, 75% of energy wastage, given that buildings use 40% of all energy used.

The primary object is that people should be at the centre of any intrusion, giving meaningful experiences. Next, the data collected must be on a central database with a holistic purpose. It has to respond to daily interactions while also plotting the long-term effect, drawing attention to where improvements can be made in future scenarios. This means learning about the buildings performance to help make data-driven solutions.

Another part of a smart building is its interoperability, which means the ability of the computer to exchange and use information seamlessly. Achieving this can often be more complex than expected. With more connected, smart devices, it is important to have a highly secured protocol to prevent potential breaches. It was also mentioned that it is crucial to consider not only what is cutting-edge today but also what will remain functional and relevant over the next decade (Cala-or 2023).

Off-Grid

Grids in energy terms are the conduits that get power from the source, whether a hydro-station or a nuclear power plant, to the end-user. They are notoriously leaky and need careful management. This ranges from 2.03% in Singapore (best) to 71.03% in Togo (worst), with USA being 5.91% and UK at 8.35% (rated as the transmission and distribution lost as a percentage of output at source (IEA 2014). By contrast in 2016, 1.4 billion people worldwide were not connected to an electricity grid (Overland 2016).

So, local sourcing has a benefit and increasingly this is happening, whether it is a solar panel in South Sudan or a microgrid in Brooklyn, new methods are prevailing.

Microgrids are localised groupings of distributed energy sources, (be it solar, wind, hydro or biomass), together with energy storage or backup generation using local management tools (Hawken 2017).

Systems can operate as standalone or can plug into a larger grid as needed, either uploading excess power or supplementing local power surges. They are nimble and efficient and as renewables and batteries improve, they will become more reliable and can work in emergency situations. Also, the use of local supply to service local demand reduces the energy lost in transmission and distribution, as noted above. In general, there is a reluctance to promote them by electricity boards as they are seen as competition to their monopolies.

Brooklyn Microgird is a community network, sharing excess energy locally, rather than uploading surplus energy to the national grid. This means that it benefits the community both environmentally and economically (Brooklyn Energy 2019). Its emphasis is on purchasing locally sourced power, changing the role of the consumer to that of a prosumer (Prentice 2018). It works quite simply with a request for electricity, which is verified by the network, approved, recorded and entered into a block of data added to the chain. Once the transaction is complete the energy is delivered.

It uses a TransActive Grid smart metering which is a meter installed in each user's house, beside to the distribution box and the domestic fuse box. If a household produces an energy surplus, either through solar power or whatever, the demand is calculated through a token system with no intermediary required. Buyers and sellers use the app to specify their preferences at what price and with whom.

Using blockchain technology, LO3 developed 'Exergy', a permissioned data platform that creates localised marketplaces for transacting energy across existing grid infrastructures through peer-to-peer (P2P) prosumers. Through the micro-grid, it is transacted locally, creating more efficient, resilient and sustainable communities. It has a distributed system operator (DSO), which has access to building management systems, and using price as a proxy it manages energy use, load balancing demand response at negotiated rates. When an electric vehicle has a surplus of energy, it can be available for purchase on the network. The micro-grid can also act as a back-up in power outages.

Servicing

Anything as a Service (*XaaS*) is something that is licensed on a subscription basis, providing an on-demand cloud service. It is something being presented to the user as a service. This 'X' can be interchanged for many differing letters, each identifying a unique IT domain. These include 'A'; API's, 'B'; Big Data, 'K'; Knowledge, 'M'; Mobility, and

'P'; Platform, to name just a few. Essentially, they are all cloud services used as and when needed. Some see them as expensive, as it is a repeating subscription, whereas buying software was a capital one-off cost, but when critical mass is reached, it enters a new world of being a true indispensable service, or so, should we hope.

Most importantly, the service is not owned, it is not bought off-the-shelf software or finished a custom-made bespoke item, that is history before it is even first deployed. The benefit to the user is that it is always up-to-date, used only when needed and always available no matter where the user is (a benefit of cloud computing).

Furthermore, they can cache numerous differing sources, to give a complete fuller picture, not dependent on a single source or stream, and this is where their majesty appears especially in mobility. Currently, a person might have a monthly commuter ticket for the train, buy one-off tickets for the bus, pay for taxis when used, rent a city bicycle or an electronic scouter, all separately. Imaging an App allowing all these things from on single central source on a smartphone.

Image credits amassing points to hire a Maserati, or a drone-taxi to the airport, or even air travel to other jurisdictions, where it was usable in New York or Boston (England), and a truly new paradigm emerges. Imagine the push/pull possibilities and the concept explodes. Differing pay plans would allow differing scopes to suit the users' requirements.

CHAPTER 10

Urban Resilience

Urban resilience is, in my opinion, a strange phenomenon. It was challenged once at a workshop with the argument: 'Why are we fixing the fallout, and not the cause?' . . . and it is a strong argument to resist, the tenet being that without addressing the cause, it leads to a bottomless pit. But it is here, and promoted by the Dutch, who live below the waterline, but manage the problem with dikes and *slusemasters* who monitor and adjust water levels as and when necessary.

It has also traipsed its way across to the United States, where it became forbidden to use the word *climate change* (in denial, some would say), meaning people had to resort to *resilience* to address the issue as a work-around. I was first made aware of this by visiting a Danish Engineering Consultancy in Washington, DC. This is their method to address rising sea levels and erratic weather. There is an agency to respond to natural disasters called the Federal Emergency Management Agency (FEMA). It is their mandate to be first responders mitigating all hazards.

Transforming the Construction Industry with Blockchain: Enhancing Efficiency, Transparency, and Collaboration, First Edition. James Harty.

'In August 2017, President Trump rescinded an executive order signed by President Barack Obama that required consideration of climate science in the design of federally funded projects. In some cases, that had meant mandatory elevation of buildings in flood-prone areas. Then in March, FEMA released a four-year strategic plan that stripped away previous mentions of climate change and sea-level rise' (Sack and Schwartz 2018).

Prime in this directive is replacing infrastructure after hurricanes such as Katrina and Sandy. When Obama was president, a corollary was added to assess replacing *like-with-like* with replacing like with better positioning so that a project would be rebuilt on higher ground, or in better protected place rather than risk a reoccurrence at a later date. The next president rescinded that order almost immediately on entering office. The goal seems to be that if a school or penitentiary is rebuilt and succumbs to the same fate at a later date, that in rebuilding, makes a tidy profit for the constructors, again, and again, ad-lib and fade.

Horses in New York

The following is a cursory tale about a perceived epidemic issue where the solution is not seen in advance, but in hindsight, it is so obvious. Blockchain and the issues it will resolve are on a similar precipice. Back in 1898, there was a conference in New York, where delegates from across the globe met for an urban conference. The issue of the day was horse manure in an urban context. It was unsanitary, a nuisance and a problem as urban developments expanded at great pace. Two decades later the automobile vanquished this planning nightmare and was hailed as an environmental saviour (Morris 2007).

In a question by Stephen Fry to QI panel on BBC (Clare Balding, Dara O Briain, Jimmy Carr and Alan Davies) (Fry 2020) Fry asks:

'How did the horses of New York City kill 20,000 people, in the year 1900?'

The ensuing dialogue, was the following:

'(In) London and other places, taxis and buses were all pulled by horses. And there were, in London alone, 50,000 horses just in the public transport system. And each one of those produces an enormous amount of poo (manure). New York City (alone produced) - 2.5 million pounds (1.1m kg) of it, every day. It was becoming an epidemic problem.

Not only was there that problem, they (the horses) were also dying. About 41 a day (NY), on average, died while working in the streets and they preferred to leave them to putrefy, because they were easier to carve up and destroy, so what we're talking about is huge quantities of manure. I mean absolutely epic, gigantic quantities, which were vectors for all kinds of diseases.

But, basically, by the time we're talking about, traffic was much more dangerous than cars, with horses, because horses themselves can bolt and drag people off with them and trample them. The noise in the city was unbelievable. The iron hooves on the cobbles was almost unbearable. You could never have a conversation on the street. And this poo that was transmitting typhus and typhoid and cholera and goodness knows what else, all kinds of unpleasant things.

And what it is that the motor car - seems peculiar to us - was an environmental saviour. It made the traffic safer, better, less smelly, faster. It was just like the answer to the city's prayers.

(Did Jeremy Clarkson put you up to this?)

. . . It's about seven times more dangerous to have horses in the city than the car, just statistically speaking'.

Before the advent of automobiles, the equine (horse) trade had a plethora of support industries supplying and serving the splendid beast. Saddlery was a bespoke industry supplying handcrafted saddles to all and sundry, covering both gentry and everyone who needed a horse. There were bridleries and blacksmiths, all in service, and ultimately hitching posts for stagecoaches to reload horsepower and replenish travel weary passengers.

All changed with the advent of the all-conquering automobile, the leather workers moved on to factory assemblies producing leather seating and trims. Blacksmiths diversified into metal works. Hitching posts became petrol stations and roadside eateries. The ultimate insult, Bugatti reduced the symbol of the horseshoe influence, into the shape of its radiator grill. In 1898 delegates from across the globe met for an urban conference, where the issue of the day was horse manure in an urban context. Two decades later the automobile vanquished this planning nightmare and was hailed as an environmental saviour.

Autonomous Mobility

'The car is not the enemy . . . (it) is the inefficiency of car ownership, where the resource sits idle for hours on end, hogs precious space and more often than not, moves only one person at a time' (Heikkilä 2014).

Moreover, auto manufacturers are becoming aware that car ownership is and will decrease, with the next generation preparing to not only not own their own house, but maybe also their own wheels. Within their orbit, the manufacturers wish to increase the average usage of a car from its current 5%, to a still pathetic 7% (Rossant and Baker 2019).

By owning a car, the owner acknowledges that it loses value before the ignition key is turned in the showroom, and throughout its life will need feeding, parking, sit in static queues or be idle for up to 80% of its life outside on the driveway or in expensive city parking houses. Such a predicament is being challenged by many where we see App´s that allow cars which are available in many urban areas to be used and dropped as and when needed. So, the ownership is removed but the service is available. Electric scooters (where tolerated) and urban bicycles also play into this scenario and are popular by both young urban revellers and tourists.

In the medium term, as this becomes more mainstream, the number of vehicles on the road will decrease. This could be as little as 20%, and if it were to happen, would change the urban landscapes and the very purpose of the street. Streets would become facilities again rather than be territorially in the realm of the vehicle. They could host activities, provide spill over spaces and bring more vegetation into our cities.

In the longer term, cars might/will be replaced by drones, meaning that rubber on asphalt will disappear, providing a big reduction in carbon emissions. This brings Luc Besson's epic file 'The Fifth Element' into perspective. Recently, a car showroom in Copenhagen had a drone-taxi in its window (Figures 10.1 and 10.2). It was a two-seater, four propellor unit, and drew a lot of attention.

Fig. 10-1: Drone-taxi in a Copenhagen car showroom.

Fig. 10-2: Interior of drone-taxi.

Just as most cars today have a gadget behind the rear-view mirror, which can immobilise the engine in hazardous conditions, this vehicle can use the same technology to navigate in three dimensions. Driverless cars are also being rolled out, with legislation the only opposition now. Machine learning captures all images from all cars to build a dataset so that it can react to most situations, and this is key to the current situation.

Driving combines continuous mental risk assessment, sensory awareness and judgement adapting to extremely variable surrounding conditions (Wadhwa and Salkever 2019). Given that robots can now driver safer than humans, how long before humans are not allowed control of a car? Already, Amazon and Google can offer drone delivery services. A coffee shop in Balbriggan, Ireland delivers coffee via drones, so it is happening piecemeal, but it is happening and will continue to grow exponentially.

Contracting in General

It is not unheard of for firms to amalgamate two or more disciplines in an effort to gain an advantage in bidding and delivering work. This could be architects and quantity surveyors or the many types of consulting engineers that we find, where the purpose is to remove potential barriers or internal conflicts to the procurement process (Smyth and Pryke 2008). This also allows for competitive bidding, offering a one-stop shop for clients, smoothing and ironing out any conflicts that might accrue with differing parties. Many consulting engineering consultants are in a frenzy to hoover up smaller firms to make them the favoured stakeholder.

This offers new methods of tendering and partnering differing consortia who often find it advantageous to come together on a project-for-project basis too. Prequalification means building-up a track record in predefined core competences. There are many strategies at play and many responses.

Looking at RIBA's work stages and Plan of Work (RIBA 2007) many combinations of procurement are mentioned including fully-designed project with single stage tender or with design by contractor or specialist; design-and-build (DB) with single or two-stage tender; partnering contracts; management contracts; public–private partnerships (PPPs) and private finance initiatives (PFIs). Within these options, there are appointments and selections, input and output packages with requirements and proposals.

Depending on the size and complexity of the project, there are a number of ways that its design and construction can be undertaken (Müller 1997). First, there is the traditional contract (Design–Bid–Build, DBB), using a standard form. It usually requires the contractor to carry out the construction according to the drawings and specification drawn up by the design team, where the work is supervised by the architect. Digitalisation in this context is a simple translation to the new media with little or no new input. All parties maintain their independence and retain their core competences.

Furthermore, it gives meaning to the levels of development (LOD's), or levels of discussion as a learned colleague calls them, where there is a progression through the procurement phase from inception to hand-over.

- LOD 100 is the generic modelling phase in making a building information model. This denotes a wall, floor or roof, for example, is placed into the model occupying a space without committing to what the materials might become. It aligns to the conceptual workstage in whatever the plan of work is being used.
- LOD 200 is the project modelling phase where there is a development in the model. This means a wall might now be brick, cavity, insulation and a light loadbearing structural inner wall. The progress here is that there is a better definition of how the building is maturing over time. It is also where most design occurs as design decisions are being made to advance the project. According to Patrick MacLeamy's graph of effort/effect over time, more decisions are being made in this phase without an increase

in remuneration or fees. By default, when a door or window is placed in the model, there is a pretty good idea of how it will look in the final model. This also means that more design alterations can happen sooner in the procurement without incurring prohibitive costs.

- LOD 300 This phase culminates the design teams' final design, ready to go out to tender. By now all materials and constructions are finished, the brick has been chosen (in colour, source and aggregate), The air-cavity's size has been calculated, the insulation has been specified and the loadbearing inner leaf has been designed. It is now in a position to be priced, time-lined and the build sequenced and planned for the next phase, construction.
- LOD 350 is an interim phase where there is a handover from the design team to the contractor. Essentially, this is what the contractor now orders, both in quantities and supply-chain availability. In a perfect world, there should be no difference between this and the former phase, but as anyone who has ever built something, it is far from the truth. A certain material might not be available, or the contractor might offer a better solution, or a detail cannot be built as planned, and this is only scratching the surface.
- LOD 400 again there should be no difference in this phase too, but changes can occur here too. Increasingly, the supply chain, through sustainability, is using Environmental Product Declarations (EPDs) which can throw the cat among the pigeons. Yes, it is an appropriate impasse, and it has a knock-on effect filtering the EPD's, and this is something I will return to because how this process is managed is critical to context.
- LOD 500 is the building procured and finished ready for handover and occupation. These last three LODs in theory should be all the same but this is rarely the case. It is best illustrated if the client requests an As-built model as part of the handover. This can be expensive as it is often made independently of the procuring parties, but as many facilities managers will tell it can be totally different. One of the few As-builts delivered was the Louis Vuitton Foundation by Frank Gehry where a weekly point-cloud scan was taken and superimposed on the BIM model, where clash detection took place, high-lighting differences, which were corrected to maintain an up-to-date digital twin.

Generally, LODs have replaced drawing scales (1 : 100, 1 : 50, 1 : 5, etc.), because modelling in reality is built at a one-to-one scale, going from the general to the particular. It should also be noted that as progress towards generative design increases, raw data will circumvent this whole process, leading towards a more automated process.

Returning to the topic of contracting, the next is a fixed price contract, where the contractor agrees to construct the building as specified in the drawings and bills of quantities for an agreed sum, by an agreed date. It allows the contractor to claim additional costs for any variations to the specification. The allowance can also be claimed for an extension of time for delays beyond his control. It can be for the whole contract, a section of work or it can be applied to a unit rate, where the price is fixed but the amount of work is unknown.

This is an area where contractors are fast becoming the drivers of the digitalisation process, with the Association of General Contractors (AGC) in America recommending to their members not to bid on non-BIM work, or in the worst case to build their own model before bidding in order to have better control on estimates and processes (Young et al. 2008). This is a significant paradigm and a move that is changing the drivers of BIM adoption. In AGC's own guide to BIM implementation, they blast the first two myths that BIM is only for large projects and large contracts (Ernstrom et al. 2010). They identify the benefits in a no-nonsense style mentioning collisions detection, visual communication, fewer errors, higher reliability, better 'what-if' scenarios, better end-product for clients and users and fewer call-backs meaning lower warranty costs.

A reimbursement contract is usually used in refurbishment work where it might be difficult to assess the cost of work beforehand, in which case it can be used to reimburse the contractor for the costs, plus a fee to cover overheads and profit. Digitalisation here, beyond the previous method, would be a checks and balance means to justify the costs. But laser scanning is being used in this niche as a means of control either at commencement of the works to record the existing work or more dynamically as a regular or daily method to track the progress of the work against the virtual model.

This is very advanced but with the development of GIS technologies this will push the augmented reality aspect of things. In design and build contracts, the contractor is responsible for the design, specification and the construction. It may be on a fixed price or cost reimbursement basis, which is either negotiated or subject to tender. It is normally used for standard or repetitive building types, where the contractor has previous experience resulting in savings for the client. This is very appealing to digitalisation especially where there is duplicity of the building type with serial clients. The benefits and return on investment make this very attractive for all involved.

Its downside raises its head, when trying to keep contractors on-board after handover, beyond a snagging period. BSRIA launched a delivery process called soft landings to smooth this operational performance and to try and keep the procurers on board for the longer stretch. The idea here is that if all are contractually involved for a longer period that there would be better decision-making during design and build. This process would be rewarded of course but would expect a better practices navigation route in return. An old mindset for contractors is to run after handover, sensing that all kinds of horrible scenarios will accrue, despite the best laid-plans. So even when forced into such arrangements, often the contractor would sell the risk onto third party enterprise risk managers to avoid any consequences, nullifying the intended outcome.

Develop and construct contracts are similar where a design team is appointed to produce concept drawings prior to going out to tender. The advantage is that the developer is only dealing with one source who has sole responsibility for the project's design and construction. This means that there is an awareness of the financial commitment prior to the commencement of construction. But they too do not address the post occupancy period.

In Management Contracting, the design team specifies the building requirements and specialist sub-contractors are supervised and co-ordinated by the management contractor to carry out the construction. For this, the management contractor receives a fee, which may be fixed or a percentage of the contract cost. It is generally used on complex projects that require a short contract period, which must have flexibility for modifications during construction. Digitalisation has a huge benefit in this fast-track method but requires a highly motivated personnel and digitally competent team.

Sub-contractors can enter into contracts with the management contractor to carry out specific work, where it is the management contractor who has a contractual relationship, not the client. If any problems arise, it is the management contractor who must pursue for a remedy. Sub-contractors are normally appointed by competitive tendering based on the drawings and bills of quantities. For the contractor, the biggest advantage here is that there are very few risks, as they are guaranteed a return of costs and they do not have the problems associated with the employment of labour, like insurance and other commitments. It is important for the project manager and quantity surveyor to control costs, since the management contractor has no incentive to control costs within the cost budget, although incentives can be introduced.

For the developer work can begin as soon as the first few work packages are produced, so allowing design and construction to overlap. The interesting part here is where the model is made available to the sub-contractor meaning the contractor has been given the model. In some cases, and pilot studies, sub-contractors were at first most reluctant to engage a model, citing all kinds of excuses ranging from beyond their scope, to tried and trusted traditional methods. But having complied there was a watershed moment of 'how had they not been doing this sooner'. I will return to this with reference to methods and programmes that are both relevant and applicable to their needs and capabilities.

Construction Management is similar in most respects to management contracting except that the contracts are made with the client. The construction manager is employed to manage the construction work. This system is used mostly on large, specialist technical projects. The payback here is that the on-site phase can be carefully monitored and fine-tuned in the model, meaning that the model is as near as the built reality as possible, and is ready for hand over to the facilities managers upon completion. The real bonus here is project certainty both in terms of time and budget, and the continued relevance of the model under operations and maintenance.

Project Management is also normally used on larger developments. It is becoming increasingly popular but more importantly it is reducing the architect's influence within the construction industry. The project manager can be an organisation or an individual, who guides the client through the procurement system, appointing the construction team and controlling the project. Usually appointed on a fee basis, it means he is not dependent on the cost of the contract. This tends to ensure that the project manager works solely in the client's interest, as he earns no commission. The potential here is that the divorcing of the architect and control of the job leaves the door open for a manager of sorts.

Whether the role is filled by someone from the construction industry or comes into the industry with pure management skills is up for debate but interestingly with regard to the building information model for which there is a need for the management of sharing, integrating, tracking, and maintaining data-sets, this offers the opportunity of a new and awesome task.

Partnering covers both PPPs and PFIs, which are special relationships between contracting parties in the design/construction industry (Erkessousi 2010). They positively encourage changes to traditional adversarial relationships to more co-operative, team-based approaches. By this, they promote the achievement of mutually beneficial goals while also preventing major disputes. PPPs are often used in infrastructural project, bridges and motorways, which result in tolls which can pay back well beyond cutting even, including fees and profit.

PFI is an arrangement where public sector assets and services are acquired through private sector funding, thus reducing government/ public sector borrowing. It is a procedure where the public sector sponsors or establishes a business case strategy. In Europe, the project is then advertised in the Official Journal of European Union (OJEU). Through a prequalification process, the bidders are then shortlisted, a consortium is usually set-up specifically for the project, forming a Special Purpose Company (SPC). The contract can then be awarded.

It is usually financed by 10% from the company, with the remainder coming from the financial institutions. It too is then recouped over the next 25–30 years from tolls or service charges upon completion. There is good potential for high returns, it gives continuity of work and offers involvement in the design phase. It means that it is highly buildable, because of the makeup of the stakeholders, and it offers more control over the programme than might be possible under traditional methods. On the downside are the initial bidding costs, which can also be long. It is a very competitive market tying up many resources initially, being also quite complex and demanding. Contract terms can also be very onerous and penalising, and it usually comes at a fixed price for the contractor.

Their main features require firm commitment from top management, encouraging continual improvement while also allowing time for the benefits to emerge. The basis for these mechanisms is to break the traditional mould, which engulfs the industry, and to provide a forum where new talent can showcase their worth. They are also based on the equality of all partners with an interest in mutual profitability. In total contrast from the traditional contract, they have free and open exchanges of information. This lends itself well to BIM and opens a bigger scenario called integrated project delivery (IPD).

Not so obvious is that it tries to keep project teams together, which is done with well thought out incentive schemes. The top reason for this is that it reduces the learning curves that are otherwise necessary, while eliminating the saw tooth information drops through work phase handovers or similar. Great value is therefore placed on long-term relationships, and an environment for long-term profitability exists where overall performance can be improved.

With all parties seeking a win-win solution everyone understands that no one benefits from exploiting each other. Crucial to this process is the problem-resolving ethos that it engenders. By having mutual objectives, the door is open to consideration, as a concept, of each other's worth. Ironically, this was missing previously. Trust and openness is encouraged to openly address problems so that innovation is embraced positively. Each partner becomes aware of the other's needs, concerns, and objectives and is interested in helping to achieve this. Prime in its objectives are improved efficiency, coupled with a reduction in costs. Dependable production quality then leads to speedier construction and more certain completion time. Longer-term benefits include better continuity of workload and a more reliable flow of design information.

The shared risk has both positive and negative connotations, to which I will return. In short, the team works much better together, but in contrast, one bad apple can turn the whole barrel. With the reduction in litigation, there is the knock-on effect of lower legal costs and exposure. There is also a lessening or removal of large contingency sums, with better decision and problem-solving systems. This in turn can equate to savings of approximately 5% on project costs, 6% for clients with profits of up to 9% accruing according to construct site (Construction Site 2010).

Where problems do occur, derives largely from the fragmented nature of the construction industry, typically the low bid mentality or from corruption. At the other end of the scale are issues of intellectual property and complacency (Williams 2009).

CHAPTER 11

Case Studies

International Terminal Waterloo Station

The International Terminal at Waterloo station by Nicolas Grimshaw is for me, one of the first buildings which could only be designed digitally. It dates from 1993 and what makes it special is that in both plans, section and elevation, no two parts are the same. Grimshaw said:

> *'The site was a nightmare. It was a completely irregular shape'.*

It spans from 55 m at the top, tapering to 35 m at the end, with a pin-jointed roof structure. The roof has a telescopic main structure of essentially three elements, which could slide over each other so that it could be adapted to fit any part of the site. It covered four platforms and snaked along the existing station with its glass and steel form on a very tight plot. This lapping characterised high-tech architecture celebrating a building's structure using industrial materials.

The letter of invitation to submit a proposal stated that 'it must be one of the most exciting commissions in Europe today'. It occupied a third of his office over 5 years with a lot of creative energy and

Transforming the Construction Industry with Blockchain: Enhancing Efficiency, Transparency, and Collaboration, First Edition. James Harty.

thousands of hours. Grimshaw sees the station as a heroic railway station with the same function as a twenty-first century airport, having all elements of an airport, with custom clearance, immigration controls and security checks. It had the capacity to handle 15 million passengers per annum or up to 1,500 in four minutes.

Not withstanding the irregular site constraints, it had a lower and flatter volume because it did not have to contend with steam and smoke, which the neighbouring station had coming from earlier times. Below the platform, there was a departure lounge and an arrival concourse and a limited car park. It also had to contend with underground tunnels criss-crossing the site at a lower level. The channel tunnel trains were also longer (400 m long) than British trains and this necessitated longer platforms.

The rendered images are very much of their time with people pasted on to the images with a shortage of materials and textures. But still it received the RIBA President's Building of the Year, and the Mies van der Rohe Award for European Architecture. It encompassed 60,000 square meters. It was a very functional design, where the roof accounted for only 10% of the cost. This was because of the simple system of units which were standardised and could be produced en masse, overlapping and being adjustable on site.

There was a lot of bespoke detailing, giving it a High-Tech architectural style. Validating the design also saw many mock-ups for both the design team and erection crews. The roof had telescopic tubes that slid inside each other but was finished a year ahead of schedule because it was largely pre-fabricated. It was also designed as a parametric building long before any parametric software was available. The concept was subsequently modelled with parametric design software, Intergraph's Ship Building Application, VDS. It was the first fully computational application of parametric design.

Terminal 5, Heathrow

Terminal 5 was a herculean project not only for the exemplary superstructure itself but also for the vast infrastructure required to service it. It cost £4.3bn., and even went through 46 months just to come through the longest public enquiry in British history. The Longford River had to be rerouted, there were two new tunnels (of nine), which were directly under the terminal for the Piccadilly underground line and its station. There was a need for reversible sidings as well as heavy freight rail provisions close to the perimeter and various other connections to the satellite terminals. It is reported that the Accident Frequency Rate (AFR) was under 25% of the national level (Ferroussat 2008).

Ferroussat also pointed out, that large projects usually go wrong citing:

- London Underground with their 2-year Jubilee Line Extension delay
- That Railtrack missed their deadline to increase slow train paths on the West Coast Main Line,
- That the Millennium Dome lurched from crisis to crisis as the bankers were brought in and
- That the British Library failed to meet the challenge of the internet age.

To correct these imbalances, he claimed that processes, the organisation and more importantly behaviours should be designed to expose and manage risk, promote and motivate opportunities and address performances in all relationships. Furthermore, leaders were to recognise that change and uncertainty was the new norm and that a different outcome meant doing something, precisely that, differently.

To conduct the procurement, a special contract was expressly written called the T5 Agreement. It included a supply chain agreement as well as an execution plan, an incentive plan, insurance, risk and changes to the scope or work. Most importantly, the incentive plan crowned the works, in that John Egan wanted the project on time and to budget. Egan was head of British Aviation Authority (BAA) at the time and insisted on his fingerprint being rite large on everything.

This was achieved through the special contract (T5 Agreement), vigorous health and safety demands, high standards of quality (behavioural approach) and using milestones to apply handover pressure in the programme. The contract was a unique legal document that managed the cause and not the effect, ensured successes in a very uncertain environment and focused on managing the risk rather than circumventing litigation. There was an incentive fund to replace normal risk payments, which funded shortfalls and provided opportunities to increase profits. Finally, the project, and not the suppliers, was insured against damage to property, injury, death and professional indemnity.

Because of carefully defining responsibility, accountability and liability, the focus became delivery. Remuneration was based on reimbursable costs plus profit with a reward package for successful completion. This incentive plan encouraged exceptional performance with the focus on the issues of value and time. Value performance occurred primarily in the design phases and was measured by the value of the reward fund for each delivery team and calculated as the sum of the relevant delivery team budget less the total cost of the work of that delivery team.

The time reward applied only during the construction stages. Here, worthwhile reward payments were available to be earned for completing critical construction milestones early or on time. If the work is done on time, a third went to the contractor, a third went back to BAA and a third went into the project-wide pot that would only be paid at the end (Ferroussat 2008). There was a no blame culture meaning that if work had to be redone the fault was not apportioned to anybody, but the rewards would either be reduced or not awarded at all. This had the effect of applying a kind of peer pressure where it was in the interest of all parties not to fail, which created a place where the vertical silos of expertise were traded for viaducts of collaborative techniques. BAA took out a single premium insurance policy for all suppliers, providing one insurance plan for the main risk. The policy covered construction and Professional Indemnity (Potts 2009). The overall supply chain was pyramidal with 80 key first tier suppliers, c. 100 other first tier suppliers, with thousands of second tier and other suppliers.

While the handbook sought to lay down binding guidelines for the whole supply chain during the procurement of the facility, it also went to great lengths to be readable for all involved and understandable in its holistic approach. It clearly defined and set out the expectations for everyone. It was ambitious with two over-riding standards: how to deliver and what is actually delivered. Best practices were benchmarked, levels of performances were outlined and expectations were raised across the enterprise. Three levels were identified; 'business as usual', which was rejected as a non-starter, 'best practice' which received an amber light and 'exceptional performance' which was seen as world class, receiving the green light and setting the bar.

Their mission was to deliver an airport, through teamwork while maintaining and delivering a strong sense of personal identity or achievement for all involved. The teams ranged from client teams through to suppliers and from management to trainees, all identified as an integral part of the supply chain. Emphasis was placed on (pre-) planning the requirements and assigning the best resources to accomplish them. Team building and their environments were cherished in an environment designed to break down (legacy) barriers and divisions. In the longer term, this impacted social and non-work relationships, to which I will return when dealing with collaboration and building interdisciplinary trust.

Appropriate training was tabled as being critical. Responsibilities, and how relationships were developed, was dealt with through ingeniously defining roles and relationships as being open, questioning and non-perspective. There was a desire to match authority with responsibility, empowering people and encouraging delegation. It was essentially a framework that had not been tried before. It was also setting out the limits and parameters to which the exceptional goals could be reached. Under behaviour colleagues were to be treated as customers, personal performance was challenged, initiative and leading by example, was encouraged and pro-active positive mindsets were seen as vital. Problems were to be dealt directly, being flexible while accommodating all contributions. All of this was to be demonstrated with proper documentation and measurement.

Finally, recognition and reward were seen as great motivators, in this postmodern relationship context. This was definitely the carrot rather than the stick driving the changes. In an interview with David Ferroussat, Commercial Quality and Resource Leader at BAA, the implementation of these principles typically saw sub-contractors having to share sensitive information. An example would be if one sub-contractor could source certain materials cheaper or better than another. The set-up meant that the one who could get the best deal supplied all; even if this meant giving a competitor (outside the contract), insight to one's coveted methods and honed procedures.

Often this would lead to an impasse where there would be blatant refusals to comply (. . . more than my job's worth . . . scenarios), which could only be resolved by taking each sub-contractor's bosses upstairs, quantifying the risk and potential loss, and paying out on the information so that the 'open' dialogue could flourish. This was unheard of in the construction industry before and for such an arrangement to flourish, a longer-term relationship is required. BAA could provide this environment with the lure of further contracts, on account of the size and scope of their organisation. This is significant while a state of transition exists.

Furthermore, when work was rejected or needed to be redone, there was a no-blame culture in place. This meant one sub-contractor could not point the blame at the other, but rather both had to submit proposals to rectify the work and correct it. The quicker and sooner that these things could be accomplished the better, because such extra work was paid out of the golden egg lump sum for finishing the work on time and to date. The longer one bickered, the more of the sum was eaten away. There was no incentive in reducing the bonus.

This was further enhanced with a critical path identifier, in the form of an object, a lump of rock, called a 'milestone', which would reside with the current critical deed, like a hot potato. The quicker the owner could pass the object on, the better, as there was a kind of a peer pressure culture established, encouraging the completion of tasks successfully to further the job and come closer to nirvana, practical completion.

So much for the management structures, on the technical side Terminal 5 was procured using Autodesk Architectural Desktop (ADT), predating Revit, which meant that the fully immersive milieu of sharing models and data did not happen in large amounts, but there was heavy involvement of NavisWorks to aid this aspect of the project (Lion 2004). The 3D co-ordination used NavisWorks as a process checker to view, review, detect clashes and extract information from the model.

There was an estimated 10% saving in design time and better co-ordination. All of the above merely co-ordinated the 3D geometry, with all output being 2D (plans, sections and elevations) extractions. This was to radically change with the advent of BIM programmes such as Autodesk Revit where data could be added to the geometry. 2D extraction is often still a legal requirement but increasingly the model is gaining in stature.

Ironically, the two satellite terminals (A and B) reverted to more regular contracts. The architects Rogers Stirk Harbour + Partners (Richard Rogers) and the engineers Arup, at one point promised Egan that they would deliver the main terminal, exactly as Egan wanted, if they could use their normal methods. Egan nearly folded to their demands but held out and declined their request.

BLOX

BLOX, situated on the Brewhouse site on the waterfront of Copenhagen Harbour, is sited on one of the most challenging plots in Copenhagen. Since 1941, no less than 75 architectural proposals had been rejected. This, it is claimed, is because it is a remote part of town, in terms of attracting people, and having a purposeful function. Then came Office of Metropolitan Architecture (OMA) who mused that a successful project here would have to be an urban connector, to enhance and energise an otherwise vacant plot. The architects have called this urban connector an urban machine, with a clear awareness that it had to contain a range of functions that would create a living building (Weiss 2018).

Part of the planning permission insisted that that Orbital ring road '02' had to remain open and in use throughout the project, which was achieved more or less as laid down. The road is an artery to Copenhagen and alternatives were not available. This created many problems during construction, as it ran through the middle of the site, but they achieved it none-the-less.

Spæncom, a precast concrete company, produced about 2,500 precast elements, 2,274 hollow-core slabs and 227 walls panels for the construction. This project is an excellent example of the benefits of BIM. The whole process has been coordinated using BIM.5D. Advanced 3D models from Spæncom and other subcontractors formed the basis for a common model, which has streamlined communication between the professional groups and, among other things, resulted in 30,000 potential collisions being resolved in the planning phase instead of at the construction site (Consolis 2023).

Detailed 4D simulations of the construction schedule also created great value. The modelling tool made it possible to achieve a 19-m deep foundation pit next to the harbour basin and to build over and under a busy road, while 25,000 cars passed through the site every day (ibid).

It has a café, a fitness centre, a restaurant, along with a playground for the local kids. It hosts marketing events, sports and other activities, with a harbour promenade and two floors of 22 apartments, ensuring life after dark. All this gives it a diversity of uses, making it a city within a city. The brief, Ellen van Loon (designer) says required a mixed-use building consisting of an architectural museum, offices and parking facilities, together with affordable housing, recreational spaces creating a metropolitan zone. Anne Skovbro says it is intended as an innovative environment, cultivation the existing environment making them both stronger while contributing something new.

Rem Koolhaas in 'Delirious New York' wrote that Manhattan is a mountain range of evidence without manifesto (Koolhaas 1994). Van Loon touches on this in BLOX saying it is a crazy intersection where you are expected to get lost. Koolhaas said that beyond a certain critical mass each structure becomes a monument and here Manhattan works as a series of mega-villages (due to zoning laws) so that each building fills up with accommodation, programme, facilities, infrastructures, machineries and technologies of unprecedented originality and complexity. This is true as someone from Upper East Side rarely goes to the Battery as both have all they need within a couple of blocks. In many ways, their goal was to produce a Manhattan block here with its mixed use and diversity, where the intention is self-sustaining to draw you in and get you lost.

Leadenhall Building

Laing O'Rourke, a large contractor in the UK, when commencing the Leadenhall Building commissioned an in-house model of the building. The model is best viewed as a YouTube video which first appeared around 2006, but soon after disappeared, only to reappear towards completion of the project in 2013. It has unique features in that it depicts the site acquisition and documents all construction phases to completion. It also added all temporary works, all utilities and the critical workforce, in sequences like the double façade system being installed.

It meant that placement of all works could be clearly seen, and the order of assembly plainly shown, drawing attention to health and safety issues, and any temporary works involved. They used multidimensional BIM technology devising an innovative construction approach. Therefore, they were able to achieve early design co-ordination, needed to meet such a challenging programme (Laing and O'Rourke 2014). It was also a difficult site, in the middle of London, with restricted street access. Being 224 m high and with 83% of the works happening off-site, the logistics of site establishment and co-ordination became programme critical. The team used BIM to perfect *just-in-time* assembly.

Also, the form of the building, leading to its nickname, the Cheesegrater (due to a planning requirement not to obstruct views of St. Paul's Cathedral, and the distinctive wedge shape reminiscent of the kitchen utensil of the same name), meant that the building was constructed slightly off vertical. This required tensioning the building back to vertical every seven stories. Radio frequency identification (RFID) was used to render a data-rich environment, enhancing control and performance, making it an early example of a digital twin.

A traditional building method was out of the question so that design for manufacture and assembly (DfMA) was used. Design for manufacture means ease of manufacture of the parts, while design for assembly means designing the product for ease of assembly on site. This also allows integrated concurrent engineering (ICE).

The model was vital for site possession, showing access, crane locations, storage, site office and all transitory functions, establishing the building site and developing it as the building grew (Offsite Management School 2015). Mobile cranes were shown being lifted into place, site hoardings protected the site and the deck, allowing ingress and egress, which could be shown most visibly and effectively.

The second stage looked at the tower superstructure, showing the building's unique form and the backbone that spawned the central core in a 'T' shaped north core. It showed the mega-frame modular construction. The north core was key to meet the challenging programme.

There was much trade integration and the construction methodology, which could be demonstrated using the modular system. It could also show the gantry cranes and brace level of the outriggers which provided structural support. Once the structural steelwork was complete, a monorail and a hinged working platform could be installed to begin the installation of the cladding panels.

The platform had three main functions. First, it provided a safety net for the workers and materials from above. Second, it provided a working platform for the hoists that guided the working cage within the lift shaft below, and third, it supported the cladding monorail needed to transport the panels into position prior to securing them to the slender lift steelwork.

Cladding installation was also demonstrated both in sequence and in manpower, striving to promote health and safety. The distinctive mega-frame supported the double façade with an inner and outer surface, with a ventilated walkway between them. The safe working area could be shown, and temporary works came and went as needed. The hanger brackets were shown for mounting, followed by the mounting of the hangers themselves, complete with workforce, machinery and safety features.

The walkways could then be shown being placed. Each external cladding panel was hoisted into place and secured in location. After the outer skin had been installed the inner double-glazed panels could be fixed. The external envelop was then fitted out with blinds, controls and electrics to monitor the interior climate. All mechanical plant was then lowered into position and moved into place where needed.

While this is shown in a video, it is important to remember that the underlying model is the powerhouse driving this project. The view can be moved to wherever it is needed. The required time sequence can be played, bringing all stakeholders onboard. The client was British Land & Oxford Properties, (the real estate arm of the Ontario's Municipal Employees) and held a 50–50% share in the project. They sold the property to CC Land in 2017 for a billion pounds sterling. Cheung Chung-Kiu is a property developer engaged in the manufacturing travel baggage.

Foundation Louis Vuitton

Frank Gehry is famous for turning scribbles into fine buildings and the Foundation Louis Vuitton is a fine example. But it is worth looking at two other buildings which defined how Gehry Technologies perform. The Walt Disney Concert Hall in Los Angeles had a budget of $50 million, but materialised at £274 million, meaning it was nearly five times over budget.

Gehry has said that his position went from having the parental role at the start of the project where he was in control, to an infantile one when cost overruns threatened to scupper it. The focus, he claims, moves from the architect to the contractor. The architect has lost face in the eyes of the owner and the contractor is now seen as the saviour if the building is to be realised. He said:

> *'You do a job; - you meet a client, they hire you to do a project, and it's usually a kind of a nice love affair and so on. It's a very positive, uplifting relationship at the start, and you develop a scheme, with plans for their building and they're upbeat and happy about it'.*
>
> *Of course, they have a budget, which they tell you and a time schedule or whatever. So, you finish the design and you put it out to bid, and then it comes in over budget. That (happens), I'd say, 80% of the time. Then the construction people say just that: we know what to do - straighten out a few things - we'll get it on budget.*
>
> *Of course, the owner finds himself very confused about this, for the most part, because they don't have the extra million dollars or whatever it is, and they're on the way or they're underway, and it's very hard to stop or be sympathetic to the architect, or to the project.*
>
> *They feel betrayed, and this happens all the time, and it's an uncomfortable place to be but no matter how much work you do, an architect can't control the marketplace, or the cost in the marketplace, or the construction world; you know, it's just not possible.*
>
> *Now you can be as careful as possible about working for budgets, but I've always hated that moment, and my friends have always hated that moment and you sort of wonder is there some way out. In the Middle Ages, the architect was a master builder, they built the cathedrals, they were respected, they had a process, and it was done over centuries, so no one got the blame, (laughs). In our time, you have the Sydney Opera House where poor Jørn Utzon gets clobbered. It's a horrible story. It practically destroyed the man's life'.*

Chronologically, The Guggenheim Museum in Bilbao followed the Concert Hall. When it was going to tender, there was a concern at what price it would come in. One of the GT team suggested taking a computer and the model to Spain to introduce and bring each of the subcontractors through the model, pulling of a bill of quantities and showing them each and every element in the project. This was rare and derided by many at the time, meaning it was pretty unique in 2004. The result was they came in at 18% under budget seeing more than a fifth being knocked of the estimate.

'And so, on Bilbao, for the steel bidding, and there is not one piece of steel that's the same if you look at the steel frame, we used CATIA. We sent a team to Bilbao and spent a week training the Sub-contractors and those people bid on the construction the steel frame. They came in 18% under budget on just the steel alone. There were six bidders and the spread between them was 1%.

Now that is knockout, rare, you don't ever get that. Which means, when you show a model of a building which looks like Disney Hall to a contractor (which we did, way back), they give you a price that's out of this world. Until you say to them 'Here's a wall that we built with it, here's the drawings. Here's like how you can do it'. The guy said 'Oh, OK!' And then you get real.

And that's what happened with Bilbao and that's what happened with all our projects since then. It's not that you control the market but that you can more precisely control the process and the things that can be controlled, you control, and it has worked beautifully'.

So now, if you want to collaborate or work with Frank Gehry, you must prequalify and use their methods so that he feels back in charge in the parental role again. Essentially, the tender went from five times over budget to a fifth under by sharing the model.

The Foundation Louis Vuitton goes a step further. It was started in 2006 and completed in 2014, consisting of eleven galleries forming an *iceberg*, which is then covered with large glass *sails*, held in position by glulam timber elements. Once the sails were formed, the glass manufacturer was asked how much bend could be placed on the glazing and this added an algorithm to the programme. This was then pitched to the framing manufacturer, and this was added as well.

The programme then took the sail shape and divided it into best fit sizes. The machine then took the first frame A1 and found a best fit for it. This continued to A2 and so on until all were sized and formed, a final check then completed the exercise, which then could be sent to the manufacturers where they were made into the various elements. It was then hoisted into place and scanned.

Because of its complexity, a digital twin was necessary. The site was regularly laser scanned and the point cloud overlaid on the model. A clash detection exercise then highlighted discrepancies, and these were corrected either on-site or in the model, meaning an *as-built* model was provided at handover.

Foundation Louis Vuitton is a new museum by Frank Gehry in the Bois de Boulogne in Paris, commenced in 2006 and completed in October 2014 for $143 million. In its procurement, three models were deployed: a design model to develop the design, a contractor (real time) model fleshing out details and a high-fidelity model, blending the two former, while providing life-cycle information for the owner in the building in-use phase (Buffa and Eastman 2014), and hence a digital twin. In the same article, Andrew Witt of Gehry Technologies went on to state:

> *'One of the objectives of the project was to create a 3D process that was for everyone to be involved. People that came on board to work on it, knew in advance that 3D was a major component for the whole process, and for those that did not have a 3D capability, Gehry Technologies had the role to consult in either training them and also to support this overall cultural change. Gehry Technologies not only provided technical support, but also helped build up a relationship of trust among the teams'.*

The structure can be divided into two parts; 11 exhibition galleries, encapsulated by 'icebergs' rising up into the firmament, which were then crowned by 12 billowing glass sails, forming a transparent cloud above the solid objects, which are elegantly held in place by a masculine glulam pivotal structure so that the whole ensemble reflects on the surrounding parkland, demarking the site and forming a gateway to the environs.

The intention towards lightness through the structure and transparency through the landscape was achieved by a rigid design procurement, with a continued process of unrelenting involvement of 10 differing disciplines, numbering over 400 individuals, working collaboratively in the development of the model and the procurement process.

The process involved taking Frank Gehry's forms and capturing their shapes digitally. This then developed into building the forms into components. These were achieved by engaging with the contractors and subcontractors, to massage the forms into buildable parts. Code was written to best serve the design intentions in the most buildable fashion. The data was transferred digitally, and computer numerical control (CNC) allowed for the robotic delivery of the building elements. As each component was lifted into place, scans were taken of the building site and overlaid on to the model. Where inaccuracies manifest themselves, either the model or the site was corrected, meaning the model was up-to-date at all times and the handover model was *as-built*, rare by most standards.

Both the complex form and juxtaposition of differing elements demanded the development of all-embracing nonstandard components (of over 200 adaptive components), which were to validate key details and produce unique solutions that were both automatic and generative descriptions of the practical instances as they were encountered.

CHAPTER 12

Sustainability

Sustainability embraces three principles in order to be inclusive and encompassing, namely: the economy, society and the environment. United Nations has defined 17 sustainable development goals describing interlinked global goals. Within construction most notably would be SDG 6: clean water and sanitation; SDG 7: affordable and clean energy; SDG 9: industry, innovation and infrastructure; SDG 11: sustainable cities and communities; SDG 12: responsible consumption and production; and SDG 13: climate action.

If cement was a sovereign country, it would be the third worse carbon dioxide polluter (8%), behind China and USA. Finding an alternative is not easy, but a development in North Dakota offers hope. BioMASON is multimillion start-up investment that uses bacteria to produce sustainable concrete masonry. It uses sand, bacteria and nutrient-rich water to ferment and create calcium carbonate crystals that binds together similar to cement. Innovative initiatives like this give hope to saving the planet.

Transforming the Construction Industry with Blockchain: Enhancing Efficiency, Transparency, and Collaboration, First Edition. James Harty.

What is disruptive, fragmented and works with insanely minimal margins? As said earlier, the answer is construction, an industry which must do better. This is an obvious observation but one which is rarely broached. What to do, and how to do it, are also questions commonly asked, but the tradition of 'this is how we always do it' and 'resistance to change' loom large.

What is needed is a method to actively engage stakeholders beyond handover, and more importantly to reward such an endeavour. More often than not the contractors seek closure upon completion and handover, resorting to even selling the risk on to third parties rather than hold on to an extended engagement through a facilities management arrangement or a leasing option so that the robustness of the handed-over building is without question fit-for-purpose and can perform its intended function for its intended life.

If there is an incentive towards continued engagement, the benefit and potential is breath-taking. Key to this is performance, and key to performance is measurement. So, if you propose a building that will save 20% in energy use for the next 20 years, a method is needed to avert green wash to deliver the goods. Such a situation could mean a pay-out of 5% of that saving for each year that the building delivers, whether it is measured in energy bills or monitored and sensored on a building dashboard. A repeating paid-out dividend is an incentive not known in the industry today.

Once established, better practices prevail, and users gets the building that they need. To verify and validate such a process blockchain enters the frame. Blockchain offers a trusted framework for the data validation. It records performance, in a decentralised, immutable open source manner. Moreover, it creates a single source of truth, eliminating latent redundancy and adversarial conflict.

To implement this new environment, a smart contract is needed. It uses *if/then* structures to administer the work. They provide protocols that verify, simplify and enforce performances. Once a party executes a task or when a milestone is achieved, a payment can be triggered. Using BIM, supplies a vehicle to this end, but this means that the model needs to become a digital twin. A digital twin needs a custodian, and this will be the architectural technologist.

Automating this process will also open the industry up to integrating the Internet of Things (IoT) and radio frequency identification (RFID). These technologies will make buildings proactive throughout their life and inform right up to their demise, whether it is transformation or decommissioning when their life-cycle closes. This makes the whole process sustainable in its purist sense. Construction accounts for nearly 40% of carbon dioxide produced on this planet and the first step in reducing this amount is to be able to measure it and determine in an evidence-based method how to tackle it. Changing how we do it and rewarding it makes sense.

Blending theory with relevant practice makes for easy reading and offers the needed information where it is most relevant. This also makes it a book which can be dipped into for specific tasks, making it a good reference book.

Aside from this, I am currently involved in a blended learning platform where blockchain is involved in both the self-assessment of the user, while in real time, it is able to identify shortfalls in the user's skill set as well as being able to direct users towards certifications that might be relevant to them, offering push/pull feedback.

Circular Economies

The RIBA Plan of Work is a well-established construction standard for organising the process of managing and designing projects, including contractual matters arising therein. It reflects on procurement practices and consequences of risk management, while considering the roles and identities of all stakeholders in the design team. It is no surprise then that it has gone through some radical changes with a new version where it has applied a morphing process from the established 2007 version, through a BIM and Green interim, to the resplendent 2013 offering. Principally, this has happened, to reflect, encourage and map the industry towards a new maturity plan, slated for 2016.

All of this is addressing the UK Government's BIM Maturity Plan of 2016. The driving force here is the public sectors' desire to control expenditure in a (client perceived) out-of-control market. It is felt that to have better bookkeeping, new methodologies must be deployed, as existing ones are seen to be both expensive and wasteful, while involving all parties in too much litigation and other disingenuous measures.

The proper use of soft-landing principles also needs to be adopted and applied to projects so that operational outcomes are understood to better match design intentions. Soft landings can also bring the client into focus and offer a method of continual contact. They also open up aftercare activities to observe and fine-tune handed over facilities. Finally, they can create a mutual culture of shared risk and responsibility, which can only help build trust and collaboration across the construction spectrum.

While The 2013 RIBA Plan of Work is a very topical subject, it is only a well guessed assessment of how new work practices are mapped (RIBA 2013). There is a pressing need for continual critique and enhancement of the 2013 Plan of Work in a number of BIM related areas including soft landings. This is most relevant for architectural technologists.

The plan is a blueprint for a longer engagement between all the stakeholders in the process of constructing buildings (Sinclair 2013a), as demonstrated in the new work stage 'In Use'. Its adoption of BIM sets out a course for moving away from (antiquated) pre-industrial methods (Kristensen 2011) to a more streamlined postmodern process where elements, assemblies and mass production enter the fray, becoming the bedrock for future work (MacLeamy 2010c). Digitalisation opens the door for a matrix of analysis, performance metrics and simulation to take centre stage, instead of the trusted pencil and tracing paper, which cannot be tested.

When examining soft landings' principles as a methodology with particular outcomes, the research question must be does it go far enough? How are contractors involved beyond the design phase, and how is the model brought on to the building site? Augmented reality (AR) and handheld devices have the power to dispel much of the antagonism and ill felt criticism surrounding the plan, but the maturity plan seems to stop short here with implementation, living up to only the Level 1/2 in the maturity plan, and currently electing not to look beyond (Sinclair 2013b).

This also extends to making explicit reference to 'PAS 1192-2/3'. PAS literally means a publically available specification, referring to the standard methodology for managing the production, distribution and quality of construction information within construction sector, by and large looking across the whole supply chain (BSI 2014). What must be asked then: are soft landings delivering this overarching aspect in the supply chain through to facility management or aftercare/user involvement?

Finally, while the subject deals with developing issues related to design management and how the profession deals with the integration of new management systems, it is imperative to stress that the soft landings encompass the life cycle analysis while a second research objective of the paper is to assess the impact this has on embodied CO_2 while sensoring/monitoring the building's post occupancy. Architectural Technologists stand to benefit most in delivering this critical piece of the jigsaw, working in unison but complimentary to architects and the other stakeholders involved (Barrett 2010).

The new plan paints a welcome broader picture of how procurement is only a part of a building's life, bringing initial strategies and sustainable mindsets to the fore (RIBA 2013). It extends, gladly, this process into the everyday use of the building, with post-occupancy checklists and environmental performance data, while ultimately giving society a better end-product. This hopefully then restores certainty and replaces rhetoric with reality (Harty and Laing 2010).

It is widely accepted that making a building is a process. It goes without saying that you cannot buy a house off a shelf. No two buildings share the same site simultaneously and hence there are basic differences, such as location and orientation, just to set the scene. Granted, there are typologies and catalogues, but essentially every building is a new beginning. Not only are the physical buildings different, but the planning and the procurement involving differing parties, and their occupancy, reflect differing peoples and their cultures.

This process is a collaboration between very many stakeholders over time, expressed through phases. These phases or work stages are designed to establish tangible milestones through the process, to aid the exchange of predefined work packages in return for remuneration. Increasingly, these packages are being addressed as soft landings (BSRIA 2014), after each phase, in a bid to better inform the client and to ease the process, ensuring that all on-board are up-to-date and best informed.

This plan of the process has a major imprimatur to remove the pretenders and improve transparency. By pretenders, I mean the deadwood and the waste that has harangued construction, detracting from productivity and increasing the unwanted excesses. This includes litigation and adverse practices. For too long construction has underperformed while being extremely wasteful of precious finite resources. In fact, poor work and double work were often encouraged as this meant double payment for the defective work to be righted. There was little incentive to get it right, it was more important to minimise your risk and maximise your profit. The business was by and large fragmented, and no one sought to rectify the situation, until John Egan and Michael Latham began voicing alternative methods and practices to bring about change (Egan 1998; Latham 1994).

But it was not until a method of co-ordinating the project, in a centralised place allowed a single point of entry, acting as an all-consuming container or moreover an umbrella organisation holding everything together entered the stage. This is commonly known as building information modelling (BIM) and it has encouraged the adoption of better working methods more than any other device in recent times (Harty and Laing 2010).

The project based collaborative process introduces a relatively new term: managed collaboration, which replaces what was previously controlled by protocols and the disciplinary silos of professional bodies who addressed their own input and interfaced with other professionals in a prepared and coordinated orderly fashion. Adopting managed collaboration is timely with the emergence of BIM and IPD, as well as lean construction, allowing integrated design.

Integrated project delivery requires a professional handling, to administer it above and beyond a mere contractual document, to define the parameters' scope and to understand each stakeholder's value. Filtering who gets what does not happen automatically, managing the enterprise needs someone who knows how to dovetail with other professionals. This situation is best suited to a hands-on technologist who can leverage things and has an all-round basis in construction (Barrett 2010).

The new plan harnesses this new situation and offers improved procurement and better service, innovation and design (Sinclair 2013a). As Darwin importantly says:

> *'. . . those who learned to collaborate and improvise most effectively prevailed'.*

Collaboration is a collective intellectual function that can be a force multiplier in an effort to reach an intended objective. In a general sense, collaboration represents a device for leveraging resources (Pressman 2014). In the same context, Scott Simpson goes as far as to say collaboration is an attitude more than a process.

Coupled with BIM being a mindset (Harty and Laing 2011), IPD comes across largely as being a method (MacLeamy 2010b) while lean construction applies very much as a process (Barrett 2008). But if collaboration skills and processes tend to transcend technology and tools, then the software does not of itself cultivate meaningful engagement. The force multiplier effect is achieved by the synergy that espouses from the common effort, adding richness and depth to the project. This is all in contrast to the traditional Design–Bid–Build (DBB) method, where the architect and the contractor were natural adversaries (Pressman 2014).

On this adversarial issue, Patrick MacLeamy went as far as to say that:

'... in fact, today's architects spend about 75% of their time on non-design tasks, practicing what I (MacLeamy) call defensive architecture. As a result, design suffers from lack of attention. Not enough time is put into thoroughly vetting the design to be sure it absolutely suits the client's purposes'.

This narrative comes from his famous MacLeamy curves talk, comparing time over effort through the construction process (MacLeamy 2010b).

Procurement now encompasses the complete supply chain and its management. Previously, a design team was assembled for the commission or project at hand and disbanded upon handover. Increasingly, design decisions are reaching beyond this ringed fence, demanding that sustainability and life cycle assessments be taken into account. Having a 'Strategic' Stage together with a 'Preparation and Briefing' Stage (Stages 0 and 1) before Concept Design (Stage 2) allows for a considered assessment of the project which can then be fleshed out before the project itself gets underway (Sinclair 2013a).

Having a new 'In Use' Stage (Stage 7) at the end begins to address the facility's management, looking at maintenance protocols and operational issues for the lifetime of the building (ibid). This larger reach and fuller commitment is intended to make the design period more meaningful. Remaining in the loop for the design team, both before and after, is a commitment to serving the client and society better. Having a bigger picture, ultimately will serve better informed design decisions and improved certainty in design matters.

Carbon Emissions

Emissions, as the term suggests, are discharged from an entity over time. In construction, this refers to the cost of a facility over its life cycle, with regard to its operating costs, its waste and its idleness. What is emitted can be a gas or a radiation but, in this instance, is carbon. While carbon is synonymous with green-house gases and carbon dioxide, some clarification might be required. Carbon is emitted by a facility when it is used. Usually, this involves heating the facility, cooling it or generally generating energy while using the facility.

Energy generation can accrue from body heat, mechanical friction or the chemical combustion with oxygen better known as fire. It is the energy to perform work, so anywhere where something happens it occurs. There is also energy that is not depleted when used and this is known as renewable, such as wind or solar. This means they are naturally replenished. In all, they are many exciting forms of renewables including solar, wind, hydro, tidal, geothermal and biomass. These will be elaborated below.

Breaking down emissions, sees 75% being energy supply (in Copenhagen), with transport and buildings accounting for 10% each, while citizen behaviour is 4% and urban development is 1% (Throssell 2009). But a lot of the energy supply goes towards or is used in buildings, so the figures are a little misleading. Generally, it is accepted that buildings account for 40% of carbon emissions. Of this the lion's share would go towards heating in a Danish climate. Building regulations have addressed this with each new edition, reducing it to a sixth of the energy used after the oil crisis.

So, as energy-efficiency is reducing in recent years, attention is being focused on the materials used and the life cycle of the facility. Realdania Foundation sponsored a project in Nyborg to build six start-up houses for first-time buyers called Mini CO_2 Houses which were completed in 2014. The first five houses were given a theme as follows; The Upcycled House, The Traditional Maintenance-Free House, The Innovative Maintenance-Free House, The Adaptable House, The Carbon Quota House and the final Carbon Standard House, launched as a catalogue house, which took the best practices and proven results of the five houses and pooled them in a standard house type (Kleis 2014).

Certifying Carbon

In 1997 in Kyoto, an agreement was reached, which essentially addressed that carbon dioxide in the atmosphere was a problem. The protocol sought to reduce greenhouse gases. In all, six gases were named: carbon dioxide (CO_2), and five other lesser but equally dangerous greenhouse gases (GHGs); methane (CH_4), nitrous oxide (N_2O), hydrofluorocarbons (HFCs), perfluorocarbons (PFCs) and sulphur hexafluoride (SF_6). The aim of the protocol was to control emissions with carbon dioxide being the biggest culprit.

Each country, that signed up to it, was given a quota to reduce their footprint. They fell into three categories; Annex I Parties who have agreed to reduce their GHG emissions below their individual base year levels, Annex I Parties who have agreed to cap their GHG emissions at their base year levels and non-Annex I Parties who are not obligated by caps or Annex I Parties with an emissions cap that allows their emissions to expand above their base year levels or countries that have not ratified the Kyoto Protocol. These are essentially, developed countries who adopt the protocol and least developed countries (LDCs) who should not foot the bill but have an obligation none-the-less. There are also countries who have not declared, have withdrawn or still have not undertaken their quotas.

For those who are in, if they met their quota, all was good. If they fell short of their quota, they would be penalised and have to buy credits to the equivalent of that shortfall. Likewise, if they met their quota with a surplice, they could sell it to countries in need. So, this provides a mechanism to encourage better practices and to reward such efforts. This gives rise to trading, and this is known as carbon emission trading.

The unit of currency is the assigned amount unit (AAU), also known as a tradable 'Kyoto Unit' and it equates to a metric tonne of carbon dioxide equivalents. They are three flexible mechanisms called emissions trading, clean development mechanisms and joint implementations. The intention with these mechanisms is to try and spread the benefits and goodwill arising out of any incentives so that developed countries can reach out to lesser developed countries.

Trading Carbon Credits

Two major market-based choices exist to deal with carbon and a method to regulate emissions: carbon trading or a carbon tax. In essence, the idea is that while new innovations and working methods cannot change overnight that a more relaxed form might achieve the same results over time. For example, Poland is very reliant of its coal mining industry, it accounted for 144 million tons in 2012, providing 55% of the primary energy consumption (75% electricity). Poland is the second largest producer in Europe and ninth in the world of coal. It is generally accepted that switching to renewable sources would wreak havoc on their fragile economy.

In Denmark too, electricity production relied heavily on Polish coal in the days of DONG (Danish Oil and Natural Gas) and when the launch of carbon credits arrived, they announced a hefty hike in electricity billing, which would have to be passed on to the end user. A political decision decided to make the price of carbon cheaper than a bottle of water to offset the increase. Since then, they have done much to improve their image, dropping the name DONG and reinventing themselves as Ørested. Their advertising now makes claims such as 'Let's create a world that runs entirely on green energy'. Their webpage announces that they are responsible for a 53% reduction of CO_2 between 2006 and 2016. They also claim that whereas coal accounted for 36% of electricity usage in 2018, today it is 100% wind turbine (Ørested 2020).

Trading in Carbon Exchanges

We need to inspire demand for sustainable energy skills by providing clear learning interactions, transparency of up-skilling transactions and recognition of qualifications achieved. Today's Millennials require training and upskilling, beyond taking notes and sitting exams. They need an education that is task-based, enabling them to think for themselves and solve problems in real time.

The skills' delivery method needs to be dynamic rather than static. The European Construction Sector Observatory Reports that 85% of all EU jobs need basic digital skills. Furthermore, 8.6 million people across the EU public sector is predicted not to have the necessary skills by 2030.

There is a demand and supply skill mismatch within construction. The skills supply mechanisms are not meeting demand. This results in skill shortages. There is also a need to have ongoing upskilling. Opportunities are being missed. We are not reaching the vocationally excluded. Vocational mobility is being restricted because the workforce does not have skills' visas.

We must address the urgency of upskilling and reskilling, transform our education interface, improve the core process of skills delivery and training, leveraging of data and technology to process and fulfil customer expectations for industry needs. This removes friction from the skills interface.

The report also acknowledges that automation and the increasingly widespread utilisation of BIM is still not widely used in the construction sector and the industry needs to develop new competences and methods of working.

The report outlines that by setting ambitious goals for Europe, the energy efficiency and energy performance of building directives have driven the need for additional green energy and energy efficient construction skills.

To achieve this, a total of three to four million construction workers in Europe will need to develop their energy efficiency related skills in the sector. SMEs are usually not interested in qualifications as they are seen as an extra cost. They are not happy to employ highly qualified people as they are afraid, they will leave for better remunerated jobs. But they are interested in employing people or collaborate with other SMEs with proven capacity to solving specific tasks. In a collaborative environment the competences can only integrate if there is trust among all stakeholders.

There are five modules to achieve this:

- Define information management processes to support condition analysis.
- Know the specific activities to enable BIM to improve performance.
- Define information requirements for the technical design stage.
- Define information requirements for the construction stage.
- Detail energy management at the operative stage.

A platform can develop an open competency-based qualification scheme based on maturity levels that empower micro-learning so that learning transactions count. It creates a system for issuing, verifying and sharing micro-credentials for learning credits, creating a digital platform which allows learners to store credentials while enabling them to share them with educational institutions and employers. It revolutionises the learning process by monetising the skills and learning exchange with a system based on skills recognition rather than accreditation.

Trust is needed to perform with others. The whole notion of trust is founded on the basics of reliability, truth, and the ability to engage with others at an acceptable level of confidence, building a platform where all can feel safe and who are empowered to participate. Automating this process removes the possibility of rogue actors becoming involved and restores a level playing field for all involved. Blockchain embraces this.

Through the use of blockchain platforms users can both use the service and enjoy additional benefits by participating in the management and control of the network. Additional benefits include verification of what you have earned to control as to how it is managed and to whom it is available. While sounding irrelevant, it is crucial to how your data is shared or displayed.

CHAPTER 13

Issues

Tokenisation

NFTs (Non-Fungible Tokens, representing physical building components in the digital world) are assets that have been tokenised by a blockchain. They have unique identification codes drawn from the metadata through an encryption functions. Fungible tokens include dollar bills because they are identical. Non-fungibles being unique, means they are different, having variable values. In construction this has two consequences, the first being a reward and personally held token reflecting the standing of the owner or procurer (designer), the latter being the project design itself having an aesthetic value.

Another consequence of these is that it promotes best possible building performances, including smart contracts to incentivise best practices. This is a new paradigm, which will grow the base within design and building practices in a way not fully appreciated today. Ultimately, it becomes a kind of payoff for all stakeholders including building owners and developers.

Transforming the Construction Industry with Blockchain: Enhancing Efficiency, Transparency, and Collaboration, First Edition. James Harty.

They can be traded and exchanged for FIAT currencies, cryptocurrencies or other NFTs, depending on the market value and prices owners have pitched to be met. Tokens are minted, meaning they are assigned a unique identifier directly linked to a blockchain address. Within construction, they are relatively new. In 2022 Alterra Worldwide launched a real-estate project by issuing $100 M security tokens with a value of $1. They are owners, developers and contactors for the project. The project was named Tower 27 and in so doing created a token named 'T27 Silicoin'. Tower 27 is a 245 storey, 374 residential units (amassing 35,597 ft^2 : 3,307 m^2), in San Jose, California.

The scale of the project allows common areas to include a pool and generous patio. A typical unit includes high end and timeless materials, a mix of natural materials, including timber, concrete and stone, resulting in a refined, clean, warm, streamlined place to live. The public spaces include a large lobby, a Gusto kitchen with community rooms, business and quiet library spaces, a game's room, fitness centre, theatre and karaoke scene as well as public restrooms.

Wallets, DAPPs and Coding

Decentralised Autonomous Organisations (DAOs) contain an internal property that has a value, and it has the ability to use that property as a mechanism for rewarding certain activities. But being decentralised, it can lead to spurious, destructive or unethical scheming, where the due process can lock the organisation in cyclic going no-where exercises. Having verifiable structures can mitigate against this, and here blockchain heralds a solution.

First, by having wallets the contents and stakeholders can vouch for their worth. Currently, this can be seen in password control apps, where they protect but by giving easy access whenever and wherever. There are also crypto wallets which can securely hold any cryptocurrencies and control the management of them. Technically, they do not hold the currencies or the credentials being vouched for, but rather they are live on the blockchain, having access only through a personal key. They allow you to send, receive or spend cryptocurrencies like Bitcoin and Ethereum.

Creating them can be an expensive process, but increasingly they are open-sourced so that a gain for one is a gain for all, making the coding involved available to everyone. In the beginning coding was seen as a lucrative exercise and being new could attract large commissions to resolve. This has had a negative response to their adoption but through the emergence of open-source solutions walls are tumbling down.

To make a transaction, two things need to be in place, first a public key which is your wallet's address (i.e. Bank account number), and second your private key (your password or PIN). It works similarly as a bank, except there is no third party, looking for a commission. Banks and financial institutions are basically not so keen as it erodes their foundations. But this is a two-way street as there are many sad cases of people losing their valuables because of the non-physical nature of the whole process.

Wallets come in many guises a cold wallet would be something like a USB stick, offline. It can also be paper based where the whole process is committed to analogue methods, away from prying eyes. Hot wallets are software based, user-friendly but less secure. They can be on a desktop, on a mobile phone, or web based. Features in wallets define user authorisation, can QR code read, can handle multiple currencies, import paper notifications, have push features and contain up-to-date conversion rates. All transactions are blockchain based, managing transactions, offering payment gateways and can logout quickly if compromised.

Incentivised by the Token System

BIM can be seen as a vehicle in the journey of procurement. As a process, it automated a technique, bringing changes in how buildings are made. Initially, it was the adoption of a new technology but quickly became a method of collaboration. It also coordinated deliverables so that plans, sections and elevations could be matched and managed in a way not done before. Furthermore, time and resources come into the mix allowing more control in the process and offering the ability to make decisions on the fly. This has grown to digital twins, where the virtual model matches the finished reality so that *what-ifs* and monitoring can be exercised through the life of the facility leading to decommissioning (demolition) and transitioning (reuse).

Common Data Environments (CDEs) essentially evolve around cloud solutions, meaning there is a single updated file at all times instead of out-of-sync revised files in several situations. Managing this service can be optimised with automation and this will be controlled by applications, *de facto* blockchain solutions.

SISK examples demonstrate case studies where digital layers improve the data flows on site through QR codes and hand-held devices. This is being adopted increasingly at a smart level and will increase in all areas of site work.

Umbrella consultancies occur where consulting engineering companies amalgamate and expand to offer clients one-stop shops so that all services can be offered within one contract, reducing sub-contracts and misaligned workflows.

Fragmentation is a huge problem in construction, for many reasons, whether it is due to many players, typically many SMEs or there is a vested interest to delay or rework tasks because it can be rewarding being paid multiply times or there is a benefit of prolonging projects.

Currently, at The Copenhagen School of Design and Technology, students must generate the following deliverables. First and foremost, they must make a study time-plan. This is a personal document that gives them an overview that they have adequate time for all the documents to be supplied for the forthcoming exam. It allows them to keep a check on their study, and to react if it goes awry.

On to this canvas, the next deliverable is their procurement strategy. This involves their contract type with consideration for the consequences of their choice. This could be Design, Bid, Build (DBB); Build and Design (B&D); Integrated Project Delivery (/Light) (IPD/Light) or a raft of other contracts. This has to be defended with arguments towards the benefit accrued and the risks averted. The evaluation criteria address the client type, the design, cost and time looking at the prospects while assessing the risks.

Under client involvement in a traditional contract, there is a call about whether the client needs to keep overall control over the design professionals. The risk here addresses whether the client has the experience and skills needed to competently deliver when required. In an IPD scenario, the criteria changes to their involvement is restricted to the early design stages where a client advisor can be engaged to protect the interests of the client.

In a traditional contract, the design must be at a high level of detail in order to attract the correct form of tender. This can be challenged when the contractor become involved, if not comprehensively designed, can lead to litigious outcomes. In an IPD scenario better buildability ensues because the contractor is involved in the design. Because it is a fluid process the client can be confused with what was asked compared to what ultimately is submitted for tender.

With regard to the time factor, in traditional situations, the delivery schedule is foreseeable and manageable. Because of its linear process, there can be longer delivery times because construction only begins after the design phase. Under IPD, construction can begin during tender design, typically, the fastest delivery system. Against this procurement can be lengthy because there are more or several options to be considered, and these must be assessed.

Under cost, the construction cost is established, meaning there should be less deviation from the initial budget. But it also means there is more responsibility on the client to meet unforeseen costs. In an IPD setting, the construction cost is uncertain at commencement, which may result in higher anticipated costs. This also moves some of the risk to contractors who must step up and cover any additional costs.

In terms of overall risk, a traditional contract means that risk can be mitigated during the design phase, but if not properly done can result in the adoption of the lowest bid which can reduce the quality of the design. With IPD, the enhanced stakeholder involvement means increased risk assessment and mitigation, meaning there is limited time to assess and resolve any overall risks. So, whether the client has the experience, is tolerant to risks, has a long-term vision, with regard to this development and any future projects, will conclude that Design and Build suits best. A letter of attorney might also be furnished too.

Once these decisions have been reached the next step is to make a preliminary project budget. Having a bubble diagram of functions and having the required square meterage, price books can advise of the anticipated rough costs. This might include or, not, site purchase, the site development and any landscaping required. Now a master time frame can be drawn up with a project start date, a construction start date and a total construction time proposal. This allows a total project duration.

In order to attract qualified relevant contractors, there is a need for a request for proposal. This will include an invitation to tender, with a description of the scope of works and the method of refurbishment. Then comes instructions for tendering, a supervised site visit, mechanisms for information requests and terms of responses. Strict dates will be outlined for all tenderers, how the proposals will be assessed and when such an event will happen. Details of remuneration and contract negotiations are also outlined here.

A Design and Build contract will now be drawn up, naming the parties, the scope of work (again), the contractual basis, the contract sum, payment of such, and a binding time schedule. How sub-contractors and sub-consultants are engaged is also addressed here. In the event of delays, redress is outlined with regard to damages which are either liquidated or assessed by law. Key personnel are also named here, primarily contact personnel, together with contractor representatives as well as resolution representatives. Performance

Fig. 13-1: Student project (Tomas Bottinelli).

bonds might also be requested, and issues of insurance are also sought. It concludes with binding signatures with the day's date.

At this stage, an early life cycle assessment might be made, as is the case with the current Danish Building Regulations (DB23) (Figure 13.1). This requires assessing Environmental Product Declarations (EPDs) which can be a very onerous exercise. Once this is in place requires target values must be submitted which become binding throughout the project.

To regulate all of this, Information and Communication Technology (ICT) process manuals are required, together with ICT specifications. These are based on a common language and how the differing stakeholders will interact. Which standards are also specified here, whether industry standards will be applied, or if there will be a tailored set of methods for the project, or whether another publication's standards will be applied for innovative work. Distribution sheets are also generated here to source what goes where and when.

Once the model is built, at LOD 100 level, a quantity take-off can be performed, which confirms the decisions taken. Resource coding can also be applied to the model, allowing scheduling to be drawn from the model. This then allows an elemental cost plan to be generated using price-books, which can then generate a better project budget (Figure 13.2). This allows for a better construction timeline to be prepared. Optionally, a 4D simulation can be generated. Parallel to this a Health and Safety plan must be made and a project specification can be deduced.

Fig. 13-2: Student project (Tomas Bottinelli).

The next phase of the design is to take a building part and go into depth with how the part can be procured. This requires a construction site layout for the specified work, a work's specification, a request for proposal from a sub-contractor, a tender list, both filled out and blank, for comparative purposed in the tender award and a detailed construction time schedule for the selected work.

Structural, loadbearing and sheer wind bracing elements must be identified and live and deal loads calculated. Bending moments and deviations are also calculated. Ventilation systems with mechanical solutions are also required. Scrutinising of the project is also conducted to test the building's performance, including acoustics, thermal bridging, moisture penetration and a raft of other things building the project's buildability, sustainability and carbon footprint.

This is an intensive process for the students but as machine learning kicks in, many if not all of, these procedures will become automated. So, this begs the role of the architectural technologist and many stakeholders in the future. Owning and being able to understand this process will make them indispensable but until we get there it is an unknown quantity, but more anon.

CHAPTER

14

Vapourised

'Spot', the robotic dog/laser scanner, is increasingly appearing on building sites going about his business scanning the project as it is at that point of time and recording it into the cloud where it is overlaid onto the building information modelling (BIM) model, whereupon the progress of the project can be ascertained with a report whether the project is ahead or behind schedule. This whole process can be run and applied automatically with no human input, unless warranted by the generated reporting, and even here the process can be driven by applications, or in another word blockchain.

CDEs can also be a tremendous aid at synchronisation of data on site. Dalux, one such CDE, has a facility to deploy Sitewalk, where a person walking the site records the current state of the site or takes 360° photos around the site. Sitewalk is a Dalux app where a camera is mounted on a hardhat, not unlike a German First World War helmet, offering a reality capture feature to faster visual documentation. The camera makes a 360° image where the front mounted camera which captures 180° and a second reverse mounted camera which captures

Transforming the Construction Industry with Blockchain: Enhancing Efficiency, Transparency, and Collaboration, First Edition. James Harty.

the remaining 180°. Stitched together they provide a total image in the round. After the walk, the images are saved under today's date and the process is repeated at weekly intervals or as needed so that there is a record of the site throughout the project.

Using a tripod, photos can be taken at predetermined points around the site. The result is much the same, but in higher quality, where the progress is recorded. Now in the app, the 3D model can be uploaded, and both the model and images can be placed side-by-side in a split screen on a device or computer. Now, they can both be synchronised having the same view in both scenes. This is then locked and so as the user either moves around the model or the image the other screen replicates the motion staying sync'd in both worlds.

This allows for a track-and-trace feature to be exercised, where the progress, or lack of, can be monitored. Because there are several dates the overall progress can be examined too. Changing the image date shows the advancement happening on the project, and this is all captured on the app. The model and image can also be merged into one where the model is half-toned, meaning fit and accuracy can also be checked. Skipping through the dates a sequence can be formed showing the progress on site. An added advantage is the operatives can see where constructions are to happen in real-time, making instant impacts for all involved.

So, with the photo alignment to the 3D model, what has been designed can be checked against what has been built, simultaneously. This enhances quality control, identifying building issues and nonconformities, while minimising errors. It also records at any point-in-time when verification of what has been installed and when. There are also savings in site visits meaning it can be visited remotely.

In the longer term, it also works towards a digital twin, as it is happening in the background, and on one single platform. It is intuitive and user-friendly and can be used in Dalux Field, Box and FM. These are features in Dalux that expand the worth and value that this can bring to all stakeholders. Field is the site feature for use on site. Box is for the design team developing the project and FM is relevant in the life-cycle assessment.

There is an impact too on historical BIM (H-BIM), where the reality capture builds a baseline of models and documentation so that any deteriorations can be mapped and tracked through a historical building's life-cycle. Subsidence can be tracked, cracks can be tagged and monitored so minor discrepancies can be supervised. The transparency that this brings to the preservation of buildings is immense.

BIM for heritage assets is a relatively recent phenomenon but is gaining traction as its merits are realised. The quality of the data is crucial here, inaccurate, incomplete or uncoordinated will result in errors, which can be detrimental to a historic asset, its value and significance. Currently, all information about a historic asset is a collection of file types and media, often being handmade drawings with parallel documentation that may or may not align with each other. Depending on the age of the asset, this can be very challenging. Some of the information may be of unknown origin, superseded or incomplete, meaning there is no single source of reliable and consistent information. Also, each asset's portfolio may have been formed independently without any coordination to the form or quality of the data held.

Another facet that is relevant here too is the role Geographical Information System (GIS) plays. BIM and GIS meet each other at a crossroads which was never fully planned or envisaged. So, their interface can be precarious. Both systems are independent of each other and owe their origins to completely different worlds. GIS relates to geography and physical assets, while BIM is independent and virtual. They can be mapped on top of each other, but the introduction ends there. GIS is a surface geometry (mesh) with little information underlying it, given where the system is coming from. BIM, by definition has a right-click button allowing properties to be added.

CDEs allow for these unique universes to collide, holding many disingenuous formats together. Archival material can be loaded in most formats, including Adobe's Portable Document Format (PDFs). Photos of hand sketches on the back of cigarette boxes can also be uploaded. Conservators are involved with the identification of the asset's strategy, pertinent to the current legislation, determining appropriate options, surveying options and the degree of appropriate intervention. Contractors assess the physical intervention and handover, while the client oversees operations and maintenance.

Often the measurements of an asset can bring another level or layer into the matrix. Very few buildings, historically, are built in the same measurement system that they now find themselves in. Most typically this might be imperial and metric systems. But before the introduction of the metric system, most countries relied on disparate systems. Now there are fast relationships between both systems but intrinsically each system brings legacy impacts to the stage. A colleague once said that measuring/surveying a building in the system that it was built reveals many nuances not appreciated in alternative systems.

Scan-to-BIM allows 3D geospatial datasets to be recorded, which involves a process of creating, manipulating and placing native components into inter-relationships. These formats can also be diverse and complex, so having a container to hold them is immense. Also, stakeholders without access to modelling software often require another format to facilitate sharing, such as Construction Operations Building Information Exchange (COBie), to manage asset information. These include equipment lists, product data, warranties, spare parts and preventative maintenance schedules.

Essentially, it replaces uncoordinated paper-based handover information created by stakeholders no-longer involved. An asset management system looks at creating/requiring assets, utilising and maintaining assets and renewing/disposing assets. It also divides what is shared, what is in progress, what is published and what is archived, so that what is relevant to whom is considered at all times. Finally, it embraces a crossover from building information modelling to asset information modelling (AIM).

CHAPTER 15

Dynamics

A consumer is an end-user who benefits from a service or product that they pay for, while a prosumer is previously a consumer as mentioned, but with a new role where they become a player, if they have excess services or products, which can then be fed back into the grid, for example. They can then be remunerated for their endeavours. Blockchain can manage and monitor this operation so that it happens seamlessly in the background with no human intervention.

Generally, this is a journey which needs to be documented, from its inception to its current state of progression. Where it will become standard is when performance becomes a benchmark to be reached and promoted to improve how buildings are made and measured. Indeed, measurement is immense (in the big picture) and increasingly it will be noted and acted upon in the decision-making processes. Just as cars are sold with reference to their performance, notably Asian models establishing themselves in most markets because of their fuel efficiency.

Transforming the Construction Industry with Blockchain: Enhancing Efficiency, Transparency, and Collaboration, First Edition. James Harty.

Brooklyn

Cities will rise to accommodate the influx of city dwellers. A new infrastructure will also be required to control and govern the apparent chaos that might ensue from our perspective today. How will these flying taxis avoid each other and how will it all be marshalled.

'Learning from Las Vegas' addresses the cognitive communication between drivers and their passengers with their *locus genii*, best described as a monument or a decorated shed. In essence Venturi claimed that in order to attract attention the objective became bigger, because the attention span was shorter due to the faster passing locations (Venturi et al. 1972). This meant that large poles with a single letter or a petrol dealer logo, pin the next burger stop or pit-stop along a motorway, or that larger than life billboards entice you off the motorway to an otherwise missed opportunity, whether it is an outlet or a mall, a picturesque beauty-spot or a heritage property.

If these semiotics fail in their purpose, then the object itself must be put on steroids, becoming either a caricature of itself or a demarked monument to why it is there. This can be best seen in the SITE warehouses of the 70s which were post-modern jokes, for want of a better word. The first time seeing a building being deconstructed by the entrance being a brick corner moving out when it is open and being pushed in to close the shop is funny and smart the first time and maybe the second time, but the tenth time, the effect is wearing off.

The road networks' take that encompasses most cities today is the product of meeting the needs of the expected automobile traffic of the past century. But significantly within this period pedestrianised city centres and historic quarters have become oases, marking out a difference within this growth sector, both legitimising it and condemning it in equal measure.

Ground-up

Many years I saw BIM as a technology, another programme or a piece of software to be learned, implemented and harvested. While in essence this is still true, my focus has changed dramatically where today it is a process, a method of sharing data and a controller of risk and certainty. Initially, it was seen by many as a procurement modeller, but now it is becoming the client's financial model, the design team's construction model and the owner's facilities management model, all bound inextricably together into a BIM pipe.

While previously they were three parallel systems, which at best glimpsed at each other across a board room table and at worst never fell into enemy hands, now there is a new culture growing, where how we can best use our models, reuse them and finally share them is to the fore. This is being adopted and propagated to deliver projects on time and to budget which are sustainable and accountably so.

A model is an incredible concept. Whether it is a pair of leggy pins on a catwalk or the ethereal notions of a mad scientist, it is a representation of a perception or a performance which can be paraded and tested before implementation. Fashion designers use a model to show off their designs, to mould the minds of the consumer to the next season's offerings and to build prestige. Scientists, on the other hand, use models to limit their scope, to test their findings and to resolve problems. Both have huge validity, whether it is through authorship or analysis.

In construction and architecture, it is the blueprint for the building, the embodiment of the design and the contractual currency for execution. It is the planning; the drawings, the specification, the quantities and the scheduling required to make a building a reality. But it is more. With digitalisation, a whole new panacea has opened.

The decision to build usually involves a financial institution or at least a financial plan. What is it going to cost, what is the budget and how is it going to be paid all come into the mix. To appraise these issues a notional building is addressed where occupancy, function, location and their impact is assessed on a spread sheet, where the building's form is not part of the equation, at least not until the money is approved. The people making these decisions are usually not spatial or graphical in their prowess and any hint of form is unwelcome and ill-advised.

So, it is often represented in a bubble diagram, for want of a better word. Large bubbles represent large spaces and often are accompanied by notional areas or numbers of occupancy, and these can be overlapping or connected by lines. As described, this work is separate from and lies outside any usable model for further work. But now it can be done within a modelling programme such as Revit which can drive the process right through to procurement. Placing a massing element, the size of the site with the desired height, or placing a parametric volume which maintains the square metres floor area, room separators can be used to generate circles and ellipses (free forms).

This process happens without defining rectangular areas which can often be misread as definitive spaces so that the abstract nature of the forms can be maintained. Tags are now added, which are as simple as name and size so that schedules of accommodation are instantly available. When the correct mix is found, price books can be associated with the data and budgetary figures are determined.

This work is phased within the model as preconstruction work. This means that it can co-exist within the model proper when construction work is subsequently prepared. The benefit of having it here is that specific climatic data can be added to this conceptual form and feedback given, regarding shape, orientation, shading, heat gain, exposure and energy performance. A report can be generated containing all the above data and if several forms are tested, several reports can be generated and cross referenced in a compare and contrast fashion, giving informed comment.

Armed with this data, the next phase of procurement is better prepared. I know some would say that this is tangential to my design methods or that this cannot, or will not affect my design process, but it can inform it. Designing for large corporate clients often means devouring volumes of standards, branding and methods that the corporation has developed so that a corporate image is presented or that certain prestige is conveyed, which during the appraisal and design brief can be very time consuming.

As the design progresses, the early work is not lost and as each form becomes an entity the early data is kept and updated and reports can affirm compliance with the initially agreed proposal. For the client, this gives a greater amount of certainty to the project which can be lost in traditional procurement methods. As the project goes through Concept, Design Development and Technical Design, these can be saved into phases in the model and through filtering of the views in the model, targeted drawings can be formatted to targeted audiences.

Typically, this might mean showing existing, demolition and new proposals tailored to whom the drawings are intended. This means the client gets colour coded legends, the planners code compliant sets of drawings and the contractor location, component and assembly drawings along with the relevant documentation all generated from the model.

Whether the output is paper format (drawings) or filtered sets (views) from the model raises a new situation, the sharing of information. The former is of no interest here as it has caused no end of checking, cross checking, red lining and revisions that eats away the fee and demoralises design teams. But how is the exchange of information handled? Who owns it, who manages it and who is responsible for it regarding errors, omissions and corrections?

If the architect initially generates a generic wall of 450 mm width, does the contractor sue or request further information when it transpires that it has to be 520 mm consisting of brick, insulation and load bearing reinforced concrete later in the design? Who is responsible for detailing the design, who is getting paid for it, or when does it come into the model following the phased work stages?

Architects or those who generally start the project mean their model is a chargeable extra. Contractors or those who take over the project feel it should be free or at no charge as it is part and parcel of the procurement. Initially the call was to give it away as it would come back in spades, but this clearly has not happened.

Under Design and Build or Partnering contracts this is abated by having all involved under the same umbrella, but even here when the work goes further down the supply chain, vested interests find it very difficult to give away trade secrets or bespoke expertise to erstwhile competitors outside of this contract. So, the new benchmark that the construction sector must address is collaboration and with whom and how.

Quintessential to collaboration is the first line of the contract, that signees will not sue each other. While sounding innocent this is a major step. Methods have to be found to remunerate work at a fair rate. Competences need to be appraised and rewarded appropriately. Changes and error rectification need to be awarded to who is best placed to do the work. There has to be an incentive to complete on time and to budget. There has to be mutual respect for all in the supply chain, and this is called plain and simple, trust.

This in turn is seeing a phenomenon of capability maturity matrices (CMMs) appearing, where differing parties tabulate their competences, their bond values and their ambitions or experience, and others compare and contrast it with their own so that strategic alliances can be formed. This is not unlike speed dating, and the metaphor does not end there. These collaborations are increasingly not for a singular project either but are related to the longer term. If a team comes together and competes and completes on a hospital (say), then they will try and corner that market and capture all related work.

Comparisons can be seen in large legal firms for architects, and also in major contractor/developer firms and large consulting engineers who feel they have the momentum to carry this off. But there is room for small players too and smaller targets but this is ongoing. When it filters all the way down to sub-contractors who can take-off quantities from the model, then significant progress has been made.

Typically, these consortia comprise a design team (architect, structural and service engineer) who use or plug into the same model. From this, a surveyor or estimator can extract quantities from the model and together with a price book or work rates and material costs can price the work. Following this, a contractor or project manager can begin sequencing the work so that there is control on site with proper manning and resources.

On site, packages can be taken off so that sub-contractors can find out how much piping, cabling or materials that needs to go into the van every Monday, and to where it is destined with how much time and how much money is allocated. This does not require the sub-contractor to have expensive authoring software (like Revit) but to only have a reader (Navisworks Freedom, Tekla Viewer or Solibri, not unlike Adobe Acrobat Reader which is free to the user), meaning they can open the file and interrogate it. This is also indispensable to the project manager in accessing the data.

If a project is authored in Revit or similar, through a process of tagging data, type codes, can be allocated. Using quantity extraction programmes a classification system can be selected and applied. This can be CAWS, Omni-class or any internal in-house system. The tagging of elements in the Revit model when complete can be exported to Sigma where all the entities can be updated against a price book such as SPONS. This gives a priced bill of quantities which can be linked back to the model so that any changes are updated in both places.

After this process, the quantities can be exported to a Gant chart programme (MS Project) where all the resources come in with the current day's date. A Gant chart allows these entities to be placed either through critical path or sequentially so that a start date and practical completion date can be calculated. Currently, this is not bidirectional with the previous work. An added chart column sequences the construction timeline. Armed with the Revit model (3D), the Project chart (4D) and the Sigma schedules (5D), these three models can be imported into Navisworks (or Tekla and Solibri).

This is done by exporting the 3D model from Revit and linking it from within Navisworks to the resources and time. A timeline feature will sequence the work which can be a movie or the project manager can move the sliding timeline bar to see the progress of the work. Selecting an element brings up when it will be built and other data like delivery and storage can be accessed or noted here.

The added benefit here is that the slider bar can be forwarded to today's date and the expected work can be compared to the actual work, meaning the manager can see if the work is ahead of schedule or behind. Navisworks will also allow him to make clash detections which can be implemented earlier in the process eliminating many architect's instructions and requests for information on site.

If the above has been carried out as described, then the final virtual model should be a replica of the actual building. This is of significant relevance to the owner or whoever is responsible for running the facility. Previously, the Facilities Manager started from scratch building a maintenance model, because often what was handed over bore no relationship to what was initially proposed.

Within the model, each element has a right click properties dialogue box built up of parameters and values. Simply said a place holder is identified and a value entered so that the 'height of a door' might have the value of '2100 mm'. During the early work stages of the project, it is of no relevance to the architect who the manufacturer is, but at some point, in the process, it will become critical, typically when the contractor is placing an order. Relevant,

Although not implemented at time of writing, there is research saying that once a door is placed in a model, that it should be possible to elicit information about it for later use. Just as smart telephone use Apps (Applications) to do things, Bots (Robots) are waiting to do embedded things. Search engines are very well-advanced today. Enter a word or topic in your browser's search engine and a meaningful response is returned, based on others who made the same search relevant to you location. All this happens in the background and, without going into the algorithms, we all use it and are relatively pleased with the hit rates and responses. That's why we come back.

Imagine a Bot placed on a door going off quietly and finding all the doors that meet the requirements demanded for that door. Initially, it might only be an internal single leaf door, with 23 manufacturers that fit the bill, but by the time, it is fully commissioned it might have a gained fire rating, a sub-mastered lock and key with particular hinges, a particular type of wooden veneer, a specific model and price, with a specified life expectancy, with inbuilt inspection periods or repair schedules. This data is only relevant to those who need it but at each stage of the process there is finger-tip informed data, at the ready, awaiting selection.

All in all, there are fascinating developments happening, and they are happening at a rate of knots. Patrick MacLeamy has engaged us with his BIM, BAM, BOOM scenario where for every dollar spent in the design phase, there are $20 spent in the assembly or construction, which leads to $60 in its operation and maintenance. If clients and users are not demanding this consideration in their projects, then we are failing them. If we are not looking at sustainable issues through all phases, then we are failing ourselves.

CHAPTER 16

Skills

Digitalisation has brought many changes in construction, and this has created a need for reskilling, in order to meet the challenge. We need to inspire demand for sustainable energy skills by providing clear learning interactions, transparency of up-skilling transactions and recognition of qualifications achieved. Today's Millennials require training and upskilling.

The European Construction Sector Observatory Reports that 85% of all EU jobs need basic digital skills. Furthermore, 8.6 million people across the EU public sector is predicted not to have the necessary skills by 2030. There is a demand and supply skill mismatch. The skills supply mechanisms are not meeting demand. This results in skill shortages. Opportunities are being missed. We are not reaching the vocationally excluded. Vocational mobility is being restricted because the workforce does have skills visas.

We must address the urgency of upskilling and reskilling. We must transform our education interface to improve the core process of skills delivery and training. Leverage data and technology processes to fulfil

Transforming the Construction Industry with Blockchain: Enhancing Efficiency, Transparency, and Collaboration, First Edition. James Harty.

customer expectations and industry needs. We need to remove any friction that exists from the skill's interface.

The European Energy Efficiency and Energy Performance of Building Directive acknowledges that automation and the increasingly widespread utilisation of BIM is still not widely used in the construction sector and the industry needs to develop new competences and methods of working. The report outlines that by setting ambitious goals for Europe, the directive has driven the need for additional green energy and energy efficient construction skills. To achieve this, a total of three to four million construction workers in Europe will need to develop their skills, where SMEs are usually not interested in qualifications as they are seen as an extra cost.

They are not happy to employ highly qualified people as they are afraid, they will leave for better remunerated jobs. But they are interested in employing people or collaborate with other SMEs with proven capacity to solving specific tasks. In a collaborative environment, the competences can only integrate if there is trust among all stakeholders their energy efficiency related skills in the sector. This is a basic failing within the construction sector because it is fragmented and works on low margins, as discussed elsewhere.

In an institution such as BuildingSmart, there are five modules:

- Define information management processes to support condition analysis.
- Know the specific activities to enable BIM to improve performance.
- Define information requirements for the technical design stage.
- Define information requirements for the construction stage.
- Detail energy management at the operative stage.

Adaptive learning develops a unique education model with personal pathways. Gamification can stimulate and engage the user, which removes a lot of friction in the process.

Trust is needed to perform with others in any industry. The whole notion of trust is founded on the basics of reliability, truth and the ability to engage with others at an acceptable level of confidence, building a platform where all can feel safe and who are empowered to participate. Automating this process removes the possibility of rogue actors becoming involved and restores a level playing field for all involved. Blockchain embraces this.

Through the use of blockchain platforms, users can both use the service and enjoy additional benefits by participating in the management and control of the network. Additional benefits include verification of what you have earned to control as to how it is managed and to whom it is available. While sounding irrelevant, it is crucial to how your data is shared or displayed.

With the increasing deployment of machine learning, there is a risk of your data falling into the wrong hands, be misused or become deepfake, where you have no control of your data and possible employers cannot trust the presented data. What is needed is a method to actively engage stakeholders beyond handover, and more importantly to reward such an endeavour. If there is an incentive towards continued engagement, the benefit and potential is breathtaking. Key to this is performance, and key to performance is measurement.

So, if you propose a building that will save 20% in energy use for the next 20 years, a method is needed to avert greenwash to deliver the goods. Such a situation could render a payout of 5% of that saving for each year that the building delivers, whether it is measured in energy bills or monitored and sensored on a building dashboard. A repeating paid-out dividend is an incentive not known in the industry today.

Once established, better practices prevail, and the user gets the building required. To verify and validate such a process, blockchain enters the frame. Blockchain offers a trusted framework for the data validation. It records performance, in a decentralised, immutable open source manner. Moreover, it creates The Single Source of Truth, eliminating latent redundancy and adversarial conflict.

By doing this mutual respect and trust can be nurtured, benefits and reward can be encouraged, more collaboration comes into the innovations and decision-making. These included wind simulation, solar gains, thermal performance and daylight factor amongst others. The adjudication parameters include architecture, energy frame, the environmental impact, the collaborative process, the application of software and mutual co-operation together within groups, polished off with an eye-catching presentation, good argument and strong validity.

Knowledge

'Education is the acquisition of the art of the utilisation of knowledge' (Whitehead 1929).

But:

'. . . educational success is no longer about reproducing content knowledge, but about extrapolating from what we know and applying that knowledge to novel situations' (Schleicher 2010).

'. . . knowing is ordinarily tacit, implicit in our patterns of action and in our feel for the stuff with which we are dealing. It seems right to say that our knowing is in our action' (Schön 1987).

Dissipating knowledge is relatively easy; ensuring that it received is totally different. In education at the beginning of the semester, the class is given a programme. During the semester, each group develops their version of the project. With reviews and in the final exam, the teacher has an opportunity to look at the work and to develop it. Sometimes, the work can stagnate and needs to be refloated. This process can encompass drawn sketches and verbal dialogue. It is something Schön calls a parallel way of designing called the language of designing (Schön 1991).

In this language, the verbal and non-verbal dimensions are closely connected and is a process of disseminating knowledge from the teacher to the student. The talk is full of psychic utterances (here, this, that . . .), which the student can only interpret by observing the teacher's movements as he sketches. The protocol is divided into several phases. First is the student's presentation and the problems encountered. Next comes the teacher's appraisal and ranking or ordering of the issues involved, which is then reframed in new terms to demonstrate a valid solution. There is a period of intermediate reflection before the next steps are outlined.

Sometimes new iterations have knock-on consequences and implications, which cut across other facets of the project, requiring further attention. This process grows the student's repertoire for problem solving and complicates the problem of discovering and honouring implications. He, who finds a problem, owns the problem is a dictate I often encourage the students to honour.

Skills

As we become more technologically sophisticated, work becomes more abstract, meaning it depends on understanding and manipulation, rather than mere acquisition. Mezirow calls this essential understanding so that learners can become effective members of the future workforce, producing autonomous responsible diagnoses.

> *'Autonomy refers to the understanding, skills and disposition necessary to become critically reflective of one's own assumptions and to engage in discourse to validate one's beliefs through the experiences of others who share universal values'* (Mezirow 2009).

This process of transformative learning involves scoping a frame of reference through critical reflection of the students' assumptions. In teaching, producing this matrix requires careful assessment of the semester assignment, aligning it to the curriculum and delivering it through the lecture plan. This requires a toolbox, whether abstract or physical, which the students can apply to their project to achieve their goal and bank for future use. Increasingly this is of a digital nature.

For the educator, this is currently a major battle between teaching too many (software) programmes and not enough hands-on detailing and key junction skills. Some say we are between stools, where previously each building was a unique tailored experience needing unique tailored solutions, to prefabrication, delivering more and more finished assemblies (and removing much of the fuzzy in situ circumstances). This argument can engage both educators and students, not to mention employers, with long held beliefs and very intransigent mindsets, long into the night.

The drawing, and the detail, is still very much currency in the construction sector. It is still part of the contractual deliverables and the point of arbitration and/or litigation, despite the legal profession not being able to analyse it as normatively as they can words. So, we must teach it. Typically, a good introduction to detailing is to take a good external-wall/pitched-roof detail, and to decline its virtues and specify each element's role in the overall key junction.

The wall and the roof need to perform throughout their life, embracing load bearing capacities, climatic shield capabilities, improving or enhancing thermal comfort, providing fire protection, weather exclusion, ventilation, durability, be maintainable and provide noise attenuation, along with a host of other things, not least cost and quality.

Previously, a massive construction like a brick wall provided many of these requirements (albeit poorly in some cases). Today, sandwich constructions provide many materials and products that function individual for each of the aforementioned things and must therefore live together with proper tolerances and an appreciation of each's shortfalls or shortcomings.

This requires taking the student through a process of scrutiny. They are shown the regulatory conditions, whether it be building regulations or local plans, they are introduced to functional analysis, leading to procuring a set of requirements for each element or component, which can now lead to a range of possibilities to be assessed. Once this process is complete, the findings are evaluated which allows them to present a knowledgeable specification. As teachers, our job is to comment both on the outcomes and the solutions.

But paper is limited, it cannot return or assess performance ratings, more importantly, it cannot be considered without 'a priori' knowledge (usually provided by the teacher team, until such a point where their knowledge base has grown proficient enough for the student to operate alone). It cannot give continuous, instant feedback, to iterations about performance such as carbon footprint of life-cycle maintenance, and so it is patently flawed.

To tackle this dimension, it is necessary to show them digital means and methods of calibrating the margins of the differing options. From this, they are then better able to make better-informed decisions and this empowers their decision-making ability.

Competences

The ECTS Users' Guide (European Commission 1989) states in its glossary that competences are:

> *A dynamic combination of cognitive and metacognitive skills, knowledge and understanding, interpersonal, intellectual and practical skills, ethical values and attitudes. Fostering competences is the object of all educational programmes. Competences are developed in all course units and assessed at different stages of a programme. Some competences are subject-area related (specific to a field of study), others are generic (common to any degree course). It is normally the case that competence development proceeds in an integrated and cyclical manner throughout a programme.*

Competences (as a broad concept) embody the ability to transfer skills and knowledge to new situations. Individuals need to be able to use a wide range of tools for interacting effectively with their environment, both physical ones such as information technology and sociocultural ones such as the use of language. They need to understand such tools well enough to adapt them for their own purposes (to use tools interactively). Increasingly, in an interdependent world, individuals need to be able to engage with others and encounter people from a range of backgrounds, especially in groups. Finally, individuals need to be able to take responsibility for managing their own lives, and to situate their lives in the broader social context as well as acting autonomously (OECD 2016).

Generic Competences

Within most disciplines, 17 competences are outlined:

- The ability to work in an interdisciplinary team
- Appreciation of diversity and multiculturality
- Basic knowledge of the field of study
- Basic knowledge of the field of the profession
- Capacity for analysis and synthesis
- Capacity for applying knowledge in practice
- Capacity for generating new ideas (creativity)
- Capacity to adapt to new situations
- Capacity to learn
- Critical and self-critical abilities
- Decision-making
- Elementary computing skills
- Ethical commitment
- Interpersonal skills
- Knowledge of a second language
- Oral and written communication in your native language
- Research skills

But Marcel R. Van der Klink and Jo Boon (2002) describe Competence as a 'Fuzzy Concept', stating that it is a useful term only bridging the gap between education and job requirements. Therefore, to ensure clarity of meaning, competences should only be defined in the vocabulary of the learning outcomes described for their programme. They should express the required competence in terms of the students achieving specific programme learning outcomes or module learning outcomes. Thus, the fuzziness of competences disappears in the clarity of learning outcomes! . . . or one should hope it is so . . .

Designing is a process where something is made (Schön 1991). Sometimes, it is the final product, where there is a hands-on process. Other times, it is incomplete, where it is only a representation, to be constructed by others. There are many variables and interactions in this process, and it can produce consequences totally different to that that was intended. In these situations, the design is fluent and must adopt these changes, forming new appreciations and understandings, which results in new situations. This is an example of the design talking-back (sic), which in a good design is reflective, or as Donald Schön describes it: where the designer reflects-in-action.

Creating this environment is challenging in the classroom, it means getting the students to be able to juggle at least three balls in the air. It is a skill that we know grows with experience, but before you have the necessary experience, it needs a substantial substitute. This to a degree is my role, to praise processes that show good decision-making, to rechannel treads that might otherwise end in disaster and to help those who are chasing their tail, or running on the spot, to get started.

It means appraising their concepts whether in materials, structural methods or sustainable solutions, encouraging them with relevant references that they can either visit or look up in the library, and quickly giving feedback to support the direction in which their project is going. This hands-on phase occurs both in consultation sessions and at evaluations where their project is presented. It is a rewarding process if the group grasps the point, which empowers their project and sometimes allows it to blossom beyond expectations.

A currency of skills means that there is a common exchange and understanding of the skills available. To achieve this common position, there has to be empathy and cross fertilisation of those skills so that they interlock and position themselves in a large matrix, shoulder to shoulder with their neighbouring skills. This interdependence creates a plethora of skills which forms a quilted blanket building into a larger tapestry to aid and abet each other.

Task-based Learning

Task-based learning draws on events to map a learning method. It removes pure theory and makes the deed relatable and identifiable. As a method, it also has a greater chance of being absorbed and remembered because seeing is doing. Using task-based learning reinforces, this condition by cutting to the chase and providing tailor-made modules to suit each and every situation. It can be seen in:

- Enabling micro-learning.
- Facilitating fast upskilling.

- Enabling personal recognition.
- Offering Units of Learning Outcomes (ULO's).

This can best be seen in up-skilling levels which interact with the user offering a better initiation point, where there is feedback giving the user an assessment of their current worth. This can work over several stages. A matrix emerges defining roles that are played in the industry. Secondly, it can rate the user from beginner to expert, and can then map a course that is relevant to the user to gain a fruitful and rewarding process. A final payoff is that as the process proceeds new endgames can become available. A user might get to within a module or two of gaining a recognised accreditation, rewarding the time input and the work done at the commencement to promote a positive response. These include:

- Level 1: Self-assessment.
- Level 2: Overview of skill gaps.
- Level 3: Personal plan.
- Level 4: Identify milestones.

This develops an open competency-based qualification scheme, based on maturity levels that empower micro-learning so that learning transactions count. It creates a system for issuing, verifying and sharing micro-credentials for learning credits, creating a digital platform allowing learners to store credentials while enabling them to share them with educational institutions and employers. It revolutionises the learning process by monetising the skills and the learning exchange with a system based on skills recognition rather than accreditation.

Verifiable Credentials

Increasingly, there is a need to be able to show achievements and gain their merit. There are many cases of cheating and misrepresentation and a method to copper fasten this is to have verifiable credentials, where a body will stand over the claims you are making. Blockchain is a solution here, acting as the single source of truth and giving the user control over who can access their credentials while offering the recipient of the data a proof-of-purchase scenario.

Blockchain is the oil that greases the cogs making machines function. Building information modelling (BIM) was the ether that made collaboration and surgery happen. Back in the middle-ages, Barber surgeons tended the wounded after battle and if amputations were required, others forcefully held the incumbent down, with a stick between their teeth as limbs were sawn off. As hospitals grew in society, operating theatres initially were placed beside the mortuary, at the periphery of the complex, so that the screams could not be heard during operations, and if unsuccessful, the surgeon could continue in the morgue, to see what went wrong.

Fast forward to the invention of anaesthetics, and the whole process changed, the patient was out-for-the-count, the surgeon relaxed, and the procedure became eminently more do-able. The result made surgeons respectable, though still not recognised in equal measure (they are still referred to as mister and not doctor). Blockchain brings this same paradigm to applications and platforms and can be the *oil* that makes all interact softly with each other. It will also bring respectability and recognition to those who use it and implement it. It can be both passive and active in push/pull situations, either protecting data, verifying data or promoting data usage.

An important benefit of blockchain is that it creates a single version of the truth, thereby eliminating redundancies, outdated records and conflicts. It also allows organisations to improve trust, efficiency and the user experience without replacing legacy systems or losing existing data. Most importantly, it can validate. This is a back-end feature, which allows employees and employers to remove a painful part of the hiring process where letters and paper versions of documents must be supplied and verified.

Blockchain offers three elements. First, blockchain has a trace and traceability, a real-time method of showing where the student is on their learning path. Second, it can be a ledger, noting what a person has learned without being compromised. Lastly, it can reward such practices with a coin that the student earns for completing modules, guaranteeing evidence to would-be third parties. While sounding relatively nominal and simple, this ability is intrinsic in a method needing transparency regarding demonstrating incorruptibility and robustness that stands up to scrutiny.

Transactional Skills

The ability to navigate this plateau and know where and what is happening, is a vital aspect of this new matrix. Having a role to massage and deliver on these skills is critical to leveraging maximum results for all concerned. It requires someone who can not only master the situation but also make significant contributions to users guiding them in the right direction, to successful conclusions.

Transactional skills also align themselves to common practice so that skills, disciplines and methodologies can be brought to bear in contract formation. This minimises the risk of litigation and brings negotiating to the fore. These skills can include entity formation, start-up financing, contract negotiation and drafting, applications for tax-exemption, commercial financing, business acquisition, commercial leasing licensing and permitting, corporate governance and compliance, real estate transactions, employment agreements and development.

Alignment as Technical Body

Technical bodies are institutes that front or combine interested bodies under a single umbrella organisation. Often, they have requirements for entry and have ethics procedures for their members to follow. They also have a purpose in keeping their membership up-to-date on current trends and legislation. They can also lobby for their interests and promote their worth. Core to their existence is a skill-set that defines them and to which they can offer in the market.

Up-skilling has a requirement from the push side of the equation so that the technical body is well fronted, and the content reflects their ethos. In preparing modules, this means making relevant modules that considers the accreditation body. If they already have course curriculum, this needs to be mapped on to the modules developed. Whether there is cross fertilisation of other modules can quite easily occur. There can be commonality in modules and this needs to respect copyrights and individuals' content.

Another aspect of this is that the technical body will most certainly have offered a curriculum before and might be reluctant to change deliverables on a different platform. If the user shall pay for the course or the certificate awarded, there is a paywall in place, with vested interests. On way around, this is for the platform to be offered for free, with all that entails, and payment happens of the award of the certificate.

Digital Building Passport

Digital wallets are beginning to show their worth, in that they verify and validate what they hold so that there is no fear of cheating or fraud regarding accreditation and certification held by the owner. Previously, on getting a job offer, the candidate would have to present all certificates and written letters about their former positions, to establish their education and experience.

CHAPTER 17

Rewarding Performance

Architecture, engineering and construction (AEC) professionals have always struggled to recover the intrinsic value of their labour. Blockchain with its properties of transparency, immutability and consensus validation now offers them an opportunity to develop a 'new value proposition' to extract reward not just for their collaborative services that they have provided but also the intrinsic intangible value of their collaborative professional service over the lifecycle of a building. Blockchain can offer a method of rewarding stakeholders who procure projects that perform better with a running contract, which if it performs pays out and likewise withholds payment for non-compliance.

Increasingly, methods are being tried that seek to prolong or engage that contractual agreement, whether it is soft-landings, life-cycle assessment or facilities management. What none of these embraces is unpredictable performance. There is also no incentive to do so or a reward system to encourage better performance.

Copenhagen wants to be a zero-carbon city. Many areas are fast promoting this both in public transportation while many

Transforming the Construction Industry with Blockchain: Enhancing Efficiency, Transparency, and Collaboration, First Edition. James Harty.

private vehicles are also going electronic. But the biggest polluter is construction where we are not making parallel leaps and bounds. Up to 40% of carbon emissions come from the construction sector, but it is not being legislated or adopted in the same manner, due to lobbying and vested interests to carry on as usual. The latest Building Regulations are addressing the situation but in its current form will only begin to make an impact in many years' time.

Adrian Malleson, in the RIBA Journal, noted that the UK government's mandate for BIM meant that performance was to the fore and would reap benefits (Malleson 2016). Just as in motoring most buyers look at performance before selecting a model, but this is not the case in construction. David Miller has said that:

> *'At the moment design decisions are all about reputation. A design that is seen as good will enhance your reputation as a designer. In the future designs will be measured against performance. And that performance has a very direct effect on the financial reward you can expect from good design. From commissioning a good designer'.*

David Ross Scheer, in his book 'The Death of Drawing, Architecture in the Age of Simulation' wrote:

> *'... whereas architectural drawings exist to represent construction, architectural simulations exist to anticipate building performance'* (Scheer 2014).

Thomas Johansen wrote, from his experience during his internship in Oslo, that:

> *'Meanwhile, we are placing ever-greater demands on our built environments, and by adding multiple components to our buildings, we increase the risk and probability of errors. The ability to just delete or change something in the model is much easier than with paper drawings. The model also allows better control ...*
>
> *... Integrated Concurrent Engineering (ICE) is a relatively new design management system that has had the opportunity to mature in recent years ... It encourages an idea that all elements of a product's life-cycle can be taken into careful consideration in the early design phases. Secondly, the concept is that the preceding design activities can all be occurring concurrently.*
>
> *Beyond the design team meeting, the debate was about ... how, the model might be given to contractors. Today, contractors are demanding models in any ... form, as it provides some sort of flow across the process'* (Johansen 2015).

The fact that paper does not facilitate this process leads to a major issue. Paper sadly becomes superfluous to the process, and this has a profound consequence. David Shepherd, author of the BIM Management Handbook, in an article entitled 'Ahead of the Game' states:

> *'Where a disruptive technology emerges – and BIM is a disruptive technology – it's effects on the mainstream is not always clear. What we might be seeing is the early stages of disruption, and in those early stages, it's very difficult to know what the effects will be. We don't have the breadth of vision to see where, and for whom, the benefits of BIM will emerge'* (Malleson 2016).

Don Tapscott saw a move from a passive internet to an active one as signalling the dawn of the Blockchain where it has a value contribution. This added a new dimension to the internet and let the genie out of the bottle, in his book; 'Blockchain Revolution' (Tapscott 2016b). His most compelling reason is that it instils a lacking-until-now trust protocol in a digital world where too much spamming, identity theft, phishing, spying, zombie-farming, hacking, bullying and data-napping (unleashing ransomware on unsuspecting innocents) is rife. Trust is needed to perform with others.

The whole notion of trust is founded on the basics of reliability, truth and the ability to engage with others at an acceptable level of confidence, building a platform where all can feel safe and who are empowered to participate. Automating this process removes the possibility of rogue actors becoming involved and restores a level playing field for all involved. Blockchain embraces this. Analogously, we have reached a stage today that someone will try to steal your wallet every step of the way, from leaving the house to going to work. This would herald a major police and public response in the analogue world. But when it is digital, it goes under the radar, largely unnoticed, but this does not lessen its seriousness.

Blockchain per se re-enforces business logistics across the board for all stakeholders, especially within a supply chain (Beck et al. 2019). It minimises transaction costs and provides transaction transparency. To do this, it brings clarity to the situation. A supply chain, whether a commodity or a service, is part of a sequence to complete a contract. Through blockchain technologies, value creators, such as designers and learners, can directly display or transfer value to their clients and employers. These values are brought to the table because of blockchain. This is an intrinsic value. Without it, no value; with it, the improvement to the service brings certainty to the proposition, meaning all concerned can operate with confidence, knowing that risk is significantly reduced.

This can also be seen when tasks are completed; blockchain can host an interface that vouches for the work, releasing payment or reward, as appropriate. It can all happen seamlessly, independently to other stakeholders. This removes payment delays, ensures that deadlines are met and rewards efficient management of workloads. This can also be seen when tasks are completed; blockchain can host an interface that vouches for the work, releasing payment or reward, as appropriate. It can all happen seamlessly, independently to other stakeholders. This removes payment delays, ensures that deadlines are met and rewards efficient management of workloads.

Malachy Mathews sees the blockchain, interlaced with BIM, as heralding a new paradigm where both the Latham report, 'Constructing the team' and the Egan report 'Rethinking Construction' can be delivered (Mathews 2017). This heady mix asks the question, how! Traditional construction contracts generally are Design–Bid–Build [DBB] where the lowest or most preferential tender is awarded the work. This has been hailed as flawed for many reasons, topmost being that in order to secure the contract, the bid usually has to be the lowest, and that the only remedy to increased revenues comes from shortfalls within the contract, the documentation or the drawings. These are litigiously examined and result in requests for information (RFIs) or change orders (COs) resulting in rework, delays and poor workmanship (Egan 1998).

RFIs usually seek to clarify further information or to provide information that was not complete at the time of signing the contract. It is good practice to include in this information, the affected parties, dates, any supporting documentation, as it will form a chain of information, creating a matrix to be tracked, answered and distributed appropriately. If this process constitutes a variation, it might qualify the relevant party to an extension of time, or a claim against losses or expenses, delaying the completion date and budget if not carefully managed. It becomes a phenomenon in itself (Aibinu et al. 2018).

Performance, Promoting Better Practices

Once collaboration is fully implemented, methods will be needed to codify good collaboration and discourage poor performance. One such method could be a distributed coin awarded for good behaviour. Often, methods are required to promote and procure better practices in collaborative environments. Peer-to-peer pressure is usually sufficient to encourage good collaboration but can be difficult to map out and adequately grade.

Bitcoin earns its stripes by the mining process, making it rare and definite. Here the earning potential is the hard work shelled out in the collaborative environment. This means it has an inherent and intrinsic value, making it valuable. Mathews has coined the term AECcoin to chart this process (Mathews 2017). The proof in the pudding is how seriously it is taken. Bitcoin had its teething problems and was abused initially, with the animosity of the coin making it of interest to the underworld and drug-related groups.

CHAPTER 18

Smart Contracts

Smart contracts are one of the most fundamental and disruptive innovations (Kinnaird and Geipel 2017). There are effectively no middlemen, and it is deemed executed, once pre-defined conditions are met. They lead to faster settlements and are very accurate. Risks are lessened and costs are reduced because there is no reliance on third parties, especially with legal costs and dispute resolutions. As the Big Data aspect increases not only will the construction branch improve but also society, in general, will appreciate the scalability and maturity of the technology.

It will improve the supply chain with its track and trace capabilities, and this will improve the disparity between Levels of Development (LOD) 300–500, which in theory in a perfect world should not change, once the design is complete. Given that buildings impact 37% of primal energy use, 38% of all carbon dioxide emissions (Carroon 2010), a new mindset is required to challenge the climate emergency. Bringing BIM and the Blockchain together, enables an entirely new paradigm for building data collection leading to truly live BIM (models).

Transforming the Construction Industry with Blockchain: Enhancing Efficiency, Transparency, and Collaboration, First Edition. James Harty.

This extends the modelling to the circular economy. BIM currently integrates 3D with 4D and 5D (geometry to time to resources) and will grow to 6D and 7D (sustainability and performance). 8D addresses optimisation and this will bring Copenhagen's desire to be carbon neutral by 2025 possible.

A drawing *(per se)* cannot plot the temperature line-loss across a construction, typically in an external wall. True, it can be made to show the loss through the cross section, but it is an abstraction that must be applied to it by a third party, performed from a third-party calculation. Likewise, a key junction cannot show depreciation over time, of the robustness of the assembly, for life cycle analysis purposes.

A 1 : 20 cross section cannot show the bearing capacity of the major structural elements to make the building stand-up. Granted, to the trained eye assumptions can be made, and the drawing can act as a container displaying the work done elsewhere to size those critical members. A virtual bucket of water cannot be thrown at a detail to test its waterproofness. No, rather the trained professional uses the presentation before him or her to apply their experience, *and only their experience*, in assessing the vigour and durability of the construction or some other competence to close the chapter.

Generally, the differing professional disciplines take pride in pin-pointing their knowledge to the presentation, in red-lining the draft before them in a time-honoured method of viewing and reviewing each other's work, in order to harmonise the construction, and drive towards a consensus during the design process. But this process is fundamentally flawed, it is open to human error. It requires many reviewing sessions, and it is open to challenge and misinterpretation.

Sarah Davidson, Gleeds Property & Construction Consultancy (Malleson 2016) says:

> *'This is a much better platform to defend a design process. Tests, simulation and performance can be carried out and recorded, and there is an audit, meaning accountability is added, and certainty is increased'.*

In the same vein, David Shepherd (Malleson 2016) saw fittingly to say:

> *'So, Level 3 won't be about being paid to produce a coordinated model, it will be about demonstrating the payback, the bottom-line value, of a design. This could be about performance of the building over its life, but it could be about the time it takes to create a building too'.*

Shephard is pointing towards performance as being as important as aesthetics to the user, the client and society in the long run. This is a serious paradigm for architects and design professionals to address. Therefore, it must be asked, how well prepared they are (architects), to deliver this important requirement? Where is the drawing's role in this abyss? How can it provide the power to communicate these parameters?

In addition, Elizabeth Kavanagh, of Stride Treglown (Malleson 2016) says:

> *'Knowledge is power, but with Level 3 BIM, power will lie in the ability to effectively share information, not in the ability to hoard it. Level 3 BIM will be about sharing the gain in a project, not allocating blame'.*

Looking at current practices, construction projects have often unwittingly accepted unacceptably higher waste levels, going above and beyond a generally accepted 25%. Waste, in this context, can be understood to include redundant document production, unused materials, idle workers, reworking and many other factors (Mays 2014). Patrick Mays goes on to say:

> *'Design-Bid-Build (DBB) contracts make it difficult for owners to derive project efficiency because of a lack of transparency in business processes and cost management systems'.*

Because no-one is in position to take over the management of the life-cycle analysis of the project, Design–Bid–Operate (DBO), similar to Design & Build (D&B), contracts provide an acceptable alternative. He says here that owners can co-ordinate the work of general contractors, subcontractors, supply chains, operation and maintenance as well as all stakeholders, to better deliver projects.

Steve Lockley (Malleson 2016) says:

> *'It looks like the genie is out of the bottle. BIM will happen. It might not be because of the mandate coming into force. But having had the mandate coming means we have become world leaders in BIM'.*

A telephone book has no place in today's animated lifestyle, full of hustle and bustle. In fact, many do not know what Yellow Pages are? To find and drill through a printed list, to find a pertinent number for what (again: what is that) is an anathema to them. Would you rather have a Filofax or a smartphone (apologies to those who have never seen the former)? Drawings are analogue too, and their fate too is sealed.

Denmark has undergone increased digitisation over the past 20 years. A digitisation strategy for it public procurement was first published in 2002 and was supposed to ensure a common procedure for digitalising in public procurement (Laustsen 2021). Digitisation was seen to be a good match for construction within the public sector (Hjelholt and Schou 2019) and therefore might influence the private sector to achieve the same productivity (Erhvervsministeriet 2019).

A report, Digital Construction, was published in 2001, which was the result of the study that was carried out in the spring of the previous year. The focus area of the report was public construction, this was in the hope that public developments could promote digitisation and thereby influence other commercial builders. When you look at the people involved in the construction projects, it is the large companies that have been the fastest to implement the use of digital tools, whereas it has been slower with the small- and medium-sized enterprises.

The digitisation of construction is aimed at the entire construction value chain and therefore has a major influence on the contractor's work and performance. This showed a change in the contractor's services; from having to deliver one performance in the form of a craft, to be able to work digitally and navigate the projects with the help of digitalisation and the execution. The increasing digitisation therefore has an influence on the contractual relationship between the client and the contractors and therefore the contractual relationship between the construction parties has been challenged.

Since the standard contractual conditions, General conditions for works and supplies in construction and construction, did not address the contractor's use of digital tools or planning obligations, a working group was set up to come up with proposals for amendments to construction contracts. In 2018, a new set of agreements was presented for the reviewers, contractors and turnkey contractors.

The Standard Form of Building Contract 2018, which became the new set of agreements obligates the construction parties to be far more solution-oriented in relation to cooperation and communication and in the use of digitalisation, including digital planning tools. Digitisation has meant that contractors had to familiarise themselves with the use of digital tools both for planning and for execution. A requirement can be made in the contract, that the contractor uses a building model for execution.

One of the challenges that the construction industry has faced with digitisation is having to navigate tender documents and revisions of drawings, descriptions and building models the following is included in the tender material. Although the construction industry has been working with digitisation for a number of years, it has been overtaken by the shipping industry.

The shipping industry, at the beginning of 2002, was not the industry that focused on digitisation. In truth, much information, documents and payments were in analogue formats. In 2016, Maersk and IBM began a collaboration about shipping container tracking. It had to be made easier for both recipient and sender to follow the delivery, including document handling of customs papers and delivery declarations. The result was a blockchain platform. The platform was launched in 2021. In short, blockchain is a database for non-auditable data. Data in a blockchain cannot be overwritten or deleted, only data can be added and read.

This has meant that Mærsk received a detailed, non-editable log and revision numbering of shipping on it all their containers. As the first worldwide shipping company, Maersk has been fully digitised in handling their main service. In principle, this shipping solution that Maersk has implemented, according to their own analysis, can be used for smart contracts.

A smart contract is a digital protocol that defines, verifies and executes specific terms and conditions, and confirms when these clauses are fulfilled. So, when Maersk receives an order for delivery of an item, a smart contract can be used to execute the order, perform payment for the parties, pay customs, port charges and other financial transactions. All of this can be practiced if the shipping order and the associated documents are tied up by a blockchain. Smart contract will be able to send shipping documentation to the various authorities and notify the recipient of the order's location on the route (Laustsen 2021).

The difference between the shipping industry and the construction industry is great and the question is whether this technology can be transferred to the construction industry. Where shipping focuses on the freight process, the construction industry focuses primarily on the procurement process. But if you look at the volume of documents and revisions of documents, there is not a big difference between shipping or construction. It is a matter of structuring all the associated documents correctly. The construction industry may be able to make use of blockchain in relation to the log of project documents and version history of digital building models.

With this log, the contractor can quickly find the current and valid digital building model and the last new revision of a given drawing. If this is possible, the contractors always have the latest new digital building model and the latest new project documents and can see the changes compared to the former revision. The construction industry is therefore looking in the direction of using blockchain in the construction process and several major law firms have employees, who work in this area with the possibilities of using blockchain in combination with smart contract in contractual relationships. Danish construction has a project group, construction blockchain, which works with blockchain technology to examine whether the use of smart contract can be implemented widely in the construction industry.

If you look at the contract area, the contracts in the shipping industry are designed for the delivery of goods. In the construction industry, the contracts are complex, as construction is a complex quantity and is based on other contractual conditions. There are usually several parties involved in a construction project, i.e. the consultants for planning and the contractors for execution. They must all cooperate to get the construction done according to the client's requirements. Therefore, there may be a number of uncertainty factors in the tender material and the interpretation of it therein. Enterprise contracts must therefore be more flexible than shipping contracts.

Construction has for some time been focused on specialist contracts. This has placed great demands on the design of the tender material and construction management, which was left to the client. This was the result of a desire by the government for increased competition, in the hope of increased efficiency and productivity. The government of the day followed suit the recommendation from the experts on the use of specialist contractors, as this could increase competition among the small- and medium-sized companies. The expert group did not find the same increased efficiency and productivity in the use of general contractor and a general contract. Therefore, support for the use of specialist contractors increased than other types of contracts.

The expert group's arguments for the use of specialist contractors lay in a financial saving, since construction management and coordination responsibility remained with the client and in the hope that the public client was more competent for construction management. The problem arose when the client lacked the right skills. The use of specialist contractors has resulted in an increasing coordination problem for the client, as several of the construction cases were of a complex size. This damaged the collaboration and did not have the desired effect. The cooperation between construction management and the trade contractors has therefore not always borne fruit.

Several of the problems lay in the design of the construction contracts and in the client's requirements. Lack of clarity in the project material, which and how the digital tools were to be used, have led to several disputes. This lack of clarity and coordination problems have resulted in demands for extra work, deadline extensions, more deficiencies and general cooperation problems professional contractor between. According to the arbitration board, this has resulted in an increase in the processing of arbitration cases. Therefore, there is an increase in the use of general contractor and general contractor for public construction.

Smart Contracts, How They Offer Solutions But Also How The Legal Elements Are Against It

Ethereum is an open-sourced platform, built on the same technology, where entities enter into contractual agreements.

'Legal Tech' is disrupting the traditional operations and self-understanding of the legal profession. This transition from analogue to digital combines 3D modelling, common data environments (CDEs) with digital tools such as BIM 360 or DALUX. During the project review, all the agreed work must be registered and written down. The work could be coded and priced in the model using blockchain. This would mean that the model must be so detailed that it constitutes a digital twin in relation to the executing contracts. Currently, this is an optional extra with an expensive price tag.

The above can be done when a receipt of the delivered materials has been made, a process check of the work performed and a final check of the work and the material and it is registered in the model. Payment can be released when work is completed, and the biggest drawback here is the retention of payment and cash flows. Payment becomes a transaction that the client cannot have a veto over, as the client must commit to the process beforehand, knowing that this process automates and improves the activity in a way not seen within construction previously. This is overcome by having parties commit to the platform so that blockchain decides when monies are released, not the client. Accepting this bind means the project can stay on time and to budget.

It goes without saying that data is a commodity; therefore, it has value. AEC professionals have traditionally, in the adversarial economy been incentivised to minimise the transfer of information between parties. This is counterproductive. Blockchain has the ability to effect even smart contracts, that is to say that if there is something to be done, when it is complete, it can be appraised in real time, and payment can be made and verified.

If industry 4.0 was about the efficiency, 5.0 leads us to develop solutions and technologies that complement human capabilities while enhancing human status. This is happening in an *ad hoc* manner, in dribs and drabs without us realising it. Contracts are about two parties agreeing something for something, occasionally with third parties involved. Ever since computer scientist and cryptographer Nick Szabo dubbed programmed agreements 'smart contracts' in 1996, there has been a hype around the self-fulfilling digital agreements that should eventually abolish lawyers and revolutionise the way we make agreements.

So, says Peter Istrup, who believes that blockchain is a hoax that does not solve a real problem. He believes that the word 'smart contracts' should be dropped for the type of agreements that are currently being made on blockchains. He prefers the term *'code'* as self-executing software. He claims the problem is that confusion arises between the objective, mathematical code language that programmers speak and the subjective language that is known from the rest of society.

The Problem Area of Current Contracting

If you focus on the construction contracts and the challenges digitisation has caused, the authorities have tried to remedy the challenges, by continuously revising the ICT Executive Order. The ICT order is information and communication technology that is used to create overall guidelines for digitisation and how the developers achieve their goal through them. By creating guidelines and openness, clients can see the value in working digitally and this means that there is value creating for clients. Developers must be able to see a benefit from the use of digitisation through three parameters, increased quality in relation to price, less extra work in the form of better and clearer project material and fewer deadline extensions.

The construction would therefore be completed at the agreed time, agreed price and agreed quality. These three parameters can be seen in the light of a desire for increased productivity, therefore have The Ministry of Transport, Building and Housing initiated the implementation of a new strategy for digital construction. This strategy, like the previous ones, focuses on productivity and efficiency. The strategy contained 5 projects and 18 initiatives. One of the initiatives dealt with digitisation and contract law. The initiative was supported of the University of Copenhagen, through the research project, 'Digital Enterprise Law'.

The project aimed to clarify the use of digitisation in the construction industry from a legal perspective. The project 'Digital Enterprise

Law' described the problems of enterprise law that could arise from the use of digital tools including digital building models. The project aimed to examine what changes digitisation had brought about and what caveats there were in entering into construction contracts.

The use of digital tools should give the construction cases a better progression, but with the knowledge that more actors must using digital tools increases the complexity of the projects, this has resulted in several arbitration awards areas that deal with deficiencies, extra work and deadline extensions. When digitisation is implemented in the construction contracts, this can have great influence based on the transactional relationship between clients and contractors. It is therefore important that the construction contracts are executed so that they are equal and in accordance with the standard form of contract.

A general 'follow-or-explain-principle' has been introduced in relation to construction contracts. The 'follow-or-explain-principle' means that the client undertakes to draw up the construction contract in accordance with standard contracts. In cases where there are changes to the individual paragraphs in the standard contract, developers must come up with a plausible and legally valid explanation for the changes in question. The builder's changes in the construction contracts unfortunately go back in history. A change, customary in design of the contracts, has not yet fully been resolved.

Derogations are still used, which are not in accordance with legal practice and good contractual practice. Changes in the construction contracts are still tipped in favour of the relationship between the client and the contractor. It is therefore one important element to examine the current beliefs and practices that characterise the construction contract system in order to understand how digitisation can contribute to a more efficient practice. Can the use of the digital tools used by the shipping industry contribute to making the construction industry more efficient and therefore whether the implementation of blockchain and smart contract in construction contract relationships can change practice or not.

The Problem with Digitalisation

The problem with digitisation and construction contracts is whether digitisation has an impact on the construction contract and the process of entering into a contract. What impact then would digitisation have on the construction contract when it comes to contractual relations, transactions between construction parties and improved cooperation.

Can the implementation of the smart contract change the challenges involved in entering into a construction contract (Laustsen 2021)?

So, what happens with increased digitalisation? How can blockchain in combination with digital building models connected by smart contracts? Do the same legal functions apply in a smart contract? Can new practices be created among the construction parties? Will these practices help improve the contractor's contract terms and what impact it has on the contractor's rights in relation to the contractual conditions being tighter with possible changes to the transaction relationship?

In order to examine the problem, there must be a focus on the analysis and the methodology using a new-institutional theory with a focus on Lounsbury's and Crumley's analysis model (Lounsbury, Crumley 2007). The model must help gain a better understanding of the dynamics and mechanisms that the immediate empirical evidence does not illuminate, by looking at the profession's actors and their understanding of the changes that may come with implementation of smart contracts.

This must be seen in order to understand the construction contract and smart contract focusing on the changes to practice that may occur by implementing new practices. Laustsen's collection of empirical evidence has taken place through interviews and literature with a view to examining construction contracts and smart contracts. He would therefore in the analysis section examine the traditional construction contract and in analysis section two, the smart contract. He would use his analysis model to compare the construction contract and the smart contract. He would use his case study to examine the possibilities in relation to setting blockchain and the digital building model in play and thereby mirror the case study, in the conditions that arise applicable when using the smart contract.

Two case studies from six, in all, used pioneering projects that can have the greatest impact on smart contracts, build trust (Frostholm 2019b) and BIM partner (Frostholm 2019a). They attempted to investigate and answer the following problem statement:

- How can smart contracts be used as construction contracts, including which problems and potential that will apply in relation to current jurisprudence?
- What are the characteristics of the current practice in relation to construction contracts?
- What are the characteristics of the new practice, in relation to the smart contracts?

- What influence can smart contracts have on the existing practice and what changes will there be in connection with the implementation of the smart contract?

Their merits and implementation will now be discussed and dissected below.

'Build Trust', Implementation of Blockchain in the Tender Process

The first pioneering project in Laustsen's case study about construction blockchain is a project called 'Build Trust'. The pioneer project is fixed against traceability using blockchain technology and must be used to create transparent documentation of the version history in documents belonging to the tender phase. It was during a user test that construction blockchain carried out that it emerged among the answers that disputes between builders and advisers sadly were quite often decided by who could store and find emails best.

The communication between the construction parties and thereby the decision-making process is fragmented, and this can create challenges for the parties in the various projects (Frostholm 2019b). The reason for the challenges in communication is that currently the construction industry is primarily divided into silos, with poor data transfer between construction parties. There is a lack of communication between the parties, which is rooted in the cultural difference between the advisers, the contractors and the builders, so clearly a mindset change is required. This can be logged but without a methodology or an offer of execution then it is a sad state of affairs.

Build trust originates from the study that was done to examine the capability of five industries to integrate new technological developments. Where the result is that the construction industry has five times as many external parties as other industries, therefore there is a need to be able to create a usable traceability in relation to documentation.

The participants in build trust were from several parts of the industry. Interest organisations were represented by Molio and Danish Industry. From the consultant side there were Vilhelm Lauritzen Architects A/S, Schmidt Hammer Lassen Architects K/S. The contractors were represented by Obos Block Watne, Gk Danmark A/S and Bravida Danmark A/S. Material producer representatives ETA Danmark A/S and IT developer 3D Byggeri Danmark ApS were also at the meeting.

Externally, there were representatives from IBM and Smith Innovation, who are responsible for delivering and supporting the technical side part of the project. These representatives were all part of groups of organisations that described themselves as innovative and helping to drive change in the construction industry. They attended the meeting because they find it an interesting relation to the future digitisation of the construction industry. Thereby they supported build trust as an innovative project that was based on a technology that has not yet been implemented in the construction industry.

'Build Trust' and Blockchain

Build trust functioned to bring together all project documents on the same platform, from the design phase to the tender material and bidding, through traceability and transparency. This traceability could be used in the future to keep track of building materials in relation to receiving control, process control and final control. This should have made it easier to handle the documents in relation to handing over construction projects. More of the participants in the project observed challenges when data and information were handed over to the parties involved. They therefore recognised the need for a common database for traceability and transparency. Build trust was based on blockchain.

The purpose of blockchain was to create a database, with a whole set of clear and defined unbreakable rules. Blockchain worked by storing a desired amount of data as a block in a database. The block was provided with a stamp that registered when the data was stored and a hash that was a unique number. If data was added to the block, a new stamp was generated since the block changed, it also got a new hash which was linked to the previous block. This therefore formed the blocks into a chain, a blockchain. The block was stored in a network where users had access.

The main idea with blockchain was to create trust across actors and organisations, in relation to the exchange of data. Blockchain was automated and it reduced the risk of errors, but the risk was not completely removed, as it could contain human errors occurring in the coding of the database. There was a risk that the consensus rule would not be respected. In public networks, blockchain opens up for greater transparency, in relation to each individual block, the stored data and the transactions that have been made.

Defence lawyer Nicolaus Falk-Scheibel is concerned about the democratisation of this in relation to the control over data in a blockchain. He sees it as a challenge that the client does not have an emergency brake in relation to the democratisation, since it is the client who pays for the construction and therefore must have the last word in relation to it stored data. This is a good example of old school clamouring for the status quo.

This problem could be solved by using a closed network where the consensus protocol was coded so that no one may have the 50% majority rule. Trust was important in the construction industry, but the parties do not trust each other, as there is generally a distrust, even if a project is based on collaboration, the parties have a dual strategy, of delivering but also maximising profit while reducing risk. Therefore, through non-editable data, trust could be created across the value chain (Mathews 2017). Trust across partners was also important to get a collaboration started and thereby create a good, coordinated dynamic.

The project differed radically from normal practice in that the trust was in the use of the technology and thereby the non-editable database. It helped to create expectations for what tender material must contain. From the above, you can see that blockchain was regulated, since it operated according to certain rules and conditions. Blockchain was additionally an electronic artefact that must get users to comply with specific sanctions.

The defence lawyers Anders Vestergaard Buch and Nicolaus Falk-Scheibel stated that blockchain was a good idea and a good tool for creating databases, but they both had caveats. Blockchain was connected with a number of challenges, such as majority rules and the anonymisation of users. Defence lawyer Anders Vestergaard Buch was concerned that blockchain could not 100% verify the building components and that the codes could be falsified.

If blockchain was put on the building components in the form of QR codes, you could register the products that were transported onto the construction site, but some controlled management needed to be carried out so that the client was sure that the right products were delivered and paid for. According to the latest knowledge about blockchain, it has become safer to use blockchain since it can be governed by a consensus coding, which ensures against hacking and digital copying, therefore the defence lawyer Anders Vestergaard Buch's concern is unfounded.

The participants agreed that the dissemination of data was important in the tender phase and that blockchain could be helpful with this. This was supported in the functionality built into blockchain. Blockchain could be part of the solution in relation to the chain of custody, the reuse of knowledge in the construction value chain and a handover of data from stakeholder to stakeholder. By looking at the observations in the case study and comparing them with the collected empirical data on blockchain, it was in line with the intended use of blockchain, as audit-proof data.

Both defence lawyer Anders Vestergaard Buch and lawyer Nicolaus Falk-Scheibel saw opportunities in blockchain in relation to data management and document management on the construction site. Blockchain in build trust was governed by community logic, as it had to help create trust across the board of organisations. Additionally, it was trust that was the primary prerequisite for legitimacy, therefore it was the pragmatic legitimacy that should carry blockchain forward in relation to recognition of the construction industry, which can demonstrate that it is useful for all users.

'Build Trust' and Smart Contract

If build trust was to form the basis of the smart contract, it was important that all the documents were coded with it through blockchain and available on the same database. The challenge arose if you included the tender documents in the smart contract too early. As a result, the tender documents could not be changed, which could now cause problems for the provider, as the tender may risk being given on an incorrect basis which could affect the project in relation to the contractor's requirements, i.e. for extra work and time limit extensions. Essentially, it seemed to thwart requests for changes and alterations as these would now be excluded from the contract as not being inclusive.

By using the smart contracts, the contractor was safe, against any changes to the tender material and an offer was made on the specific tender material. A smart contract was a protocol that defined, verified and executed specific contract terms. Alexander Savelyev defined it as: '*[. . .] we define a Smart Contract as an agreement in digital form that is self-executing and self-enforcing*'. Therefore, becoming a smart contract was self-fulfilling and differed from other contracts where the fulfilment of the transactions was the task of the contracting parties. A smart contract was regulation and supported by symbolic systems for the set-up rules, laws of execution, management protocols and standard operations inscribed in the connected blockchain.

Build trust set out regulatory standards, built around artifacts, the documents in the tender and normative standards carried by relational systems, where it was the relationship between the organisations, which were linked together and thereby built trust. It was all built on trust in the database and could be connected by smart contracts. Build trust was governed by community logic, as the primary focus of blockchain was to create trust between all parties. A smart contract was pragmatically legitimised as it supported an organisational requirement against a counterparty. In so doing, the organisations acknowledge this dependency.

'Build Trust' and Future Visions

The future vision for the project was whether it would eventually be possible to implement the digital building model on the platform. Only data from the digital building model would be needed and not the full model. Data was read back to the digital building model through the use of blockchain, thereby ensuring verification and audit security of the model. There would be a need to expand the platform for it to be able to handle this information. Expansion of the project platform had to take place as the ideas increased. Work was based on an agile transformation of the platform so that it was future-proof.

The future use of build trust was to be able to supplement the client with documenting the sustainable measures taken in the project and therefore showed a reuse potential in the built-in components. To make the project useful in the construction industry, construction blockchain should focus on the fact that build trust could help promote communication and trust could be improved by using non-editable data for the exchange of information, documents and models between the construction parties. This was the most important point of the whole project, to make the industry understand that Build Trust could help bring trust back to the correct tender material and the audit-proof system.

It was important to mobilise the project participants so that through their network they could get the construction industry to use build trust as an integral part of the work. The construction industry needed to be convinced that Build Trust used non-editable databases for storage that was far more secure. The project was a niche project, but with the correct business case it could become a sensible and functional tool.

'BIM Partner', Implementation of Blockchain with BIM

The other pioneering project observed in the construction blockchain was BIM Partner. The project focused on the use of blockchain for audit protection of digital building models through a digital platform. The project should ensure the same set-up of the digital infrastructure. The platform could be used to register the client's requirements and compare them with the digital building model and assess whether the requirements had been met.

The project focused on blockchain to validate data stored in the model as agreed in, the delivery specification, client requirements and progression in the completion of the digital building model. The digital building model was stored on a digital platform where the various client requirements for the model could be placed. As in build trust, the participants were representatives from the broad part of the construction industry, and they represented all parts of the disciplinary silos.

Contrary to build trust, the stakeholders mentioned in this project were Vilhelm Lauritzen Architects A/S and HD Lab. Both actors were proponents of increasing innovation in the construction industry and were the driving forces, among other things, during construction of the VCD labs and digital platforms for information sharing.

Through the BIM partner platform, the basis of the agreement between the consultant and the client could be checked. This was done by placing blockchain on the digital building objects in the building model and extracting a protocol from which requirements had been complied with. The platform could continuously ensure that the requirements were met when revisions of the digital building model were prepared. The participants had observed that there was a discrepancy between the client's requirements and the information and data that was stored by the reviewers in the digital building model.

The platform could provide users with an update, in relation to progression, in the digital building model. The advantage was that the digital building model was logged, so you could see which changes had been made and by whom. A discrepancy had been observed between participants between the information level the digital building model should be delivered in, and the phased design model used in the architectural industry. By using blockchain in combination with the digital building model, it increased confidence that the model had been verified and controlled in accordance to given metadata used as a control.

The way blockchain was used in this project was that each actor had a copy of the database that was linked together with the other players' databases. It was all linked up in the digital building model. This meant that new blockchain was stored in the database and distributed to stakeholders copies of the digital building model. Blockchain can in this case verify the participants in the network, this is called proof of identity. To verify transactions in the network, proof of work was used.

BIM partner shifted the focus to data in the digital building model. Shifting the focus was in-line with the defence lawyer Anders Vestergaard

Buch and lawyer Nicolaus Falk-Scheibel, who had stated that the priority between the various project documents there was a shift from drawings and descriptions to a digital building model, where data was stored that could be used to formulate descriptions for the entire building and the various building parts. In order for the construction industry to see the idea of using BIM partner, the project participants would have to comply with a single strategy. Therefore, the primary focus area was the validation of the model held up against the client's requirements.

By drawing on the client's requirements, construction's blockchain could mobilise the client, they would be able to see value creation so that their requirements were met for the entire construction process. In this way, the builders could help create pragmatic legitimacy, where BIM partner was widely accepted by the industry.

Smart Contracts, the Legal Parameters and Challenges

By virtue of the fact that smart contracts were computer controlled, self-fulfilling, immutable and rigid, one cannot avoid examining this type of contract in relation to the principles of the law of obligations, including the Contracts Act. As mentioned, the law of obligations was a foundation for all agreements and should be observed. By examining the smart contract, this basically applied. The challenge was largely coming from the contract lawyers, being set in their ways.

The Contracts Act requires acceptance of offers and invalid declarations of intent. Therefore, there should be proof that the offer had been accepted primarily in the form of a signature. In relation to Chapter 1 of the Contracts Act, there was freedom of agreement, the agreement or contract could be defined as text or computer codes, this did not affect the validity of the contract. It could be a problem for the judges to read a smart contract, therefore it was appropriate to have the contract translated.

Chapter 3 of the Contracts Act applied when entering into a smart contract. Therefore, a court could overrule a smart contact if it did not comply with the overruling contract law. The smart contract was immutable, this was the security and predictability in the execution of the contract that helped to legitimise it as a useful part of the contract system. The predictability was the advantage of the smart contract but also its Achilles heel.

It was associated with disadvantages that the contract could not be changed if conditions arose in the agreement that could not be implemented or with the termination of the agreement. If errors were discovered in coding that could be misused, these could not be changed. It was therefore not possible to correct the error or additions to the contract. In relation to contract law and smart contract, contract law was normative, and governed by norms and principles within legal practice. The Contracts Act was a regulation, as it was written legislation. Both were supported by symbolic systems, laws, rules, principles and values. Enterprise law and the contract law were based on the logic of justice.

'BIM Partner' and Smart Contracts

In order for the smart contract to be used together with BIM partner and the construction blockchain, it would be that the digital building model needed to be validated. This allowed the digital building model with its blockchain content was used as part of the tender material. Blockchain could be used to validate smart contracts in relation to execution, if blockchain was coupled up to the digital building model. Therefore, it became the digital building model, together with blockchain that governed elements of a smart contract.

The digital building model was updated before the tender and so included the builder's requirements. The challenge lay-in if errors occurred in metadata. If there were errors in metadata it would lead to errors, which would be transferred to the digital building model and further into the blockchain. The error would not be detected by the quality control as it was controlled by a log file generated by the link between metadata and validated data in the digital building model. If this error was not detected by a human control, the error would be transferred to the smart contract, and this would have unpredictable consequences in relation to the project.

The smart contract was based on community logic, as it was based on trust-building by the use of non-editable data, the predictability of execution and the immutability of the chain. As a starting point, these two logics could work well together, it was just the logic of justice that was the primary and the community logic, which was the secondary one. Smart contract was legitimised by regulation, as it was managed according to rules and laws in the form of coding in the various programmes.

'BIM Partner' and Future Visions

The future of the BIM partner project will increasingly focus on the implementation of smart contracts, blockchain and the digital building model. It was discussed at the meeting whether it was possible to make another coupling other digital platforms so that these could be used for any tender and execution. The collaboration between the digital platforms could help control the execution. It was still unknown whether this interconnection could work, but solutions would be looked at. The link between blockchain and the digital building model could present challenges in relation to the number of nodes that were in the network and to the different connections blockchains had in the digital building model.

The more nodes, the greater the risk that the amount of blockchains, changes and additions of data, would become unmanageable, meaning scalability. A node was the bond that was between the different blocks in a blockchain. The advantage of using blockchain was, that one could not argue against an algorithm in relation to the placement of responsibility. Therefore, transparency was a good thing idea in relation to responsibility, because you could assess what legal relationships existed between the parties in relation to the digital building model.

The participants, in BIM partner, were positive about the link between the digital building model, blockchain and smart contracts. Smart contracts was in combination with BIM partner regulations, carried by a symbolic system, as the contract prescribed rules and laws in relation to the execution of the content. BIM Partner was governed by the community logic, as it was based on the trust algorithm that was in the blockchain. BIM partner was legally legitimised, as it was controlled by algorithmic laws and regulations encoded in the system.

Summary of Analysis

The connection between the construction blockchain and smart contracts is complicated. Both pioneering projects can, as starting point be used in combination with smart contracts, because build trust focused on the tender material and BIM partner focused on the digital building model. The challenge was that the smart contracts could not control non-digital activities. It was a limitation that the smart contracts had and one which cannot be overcome without. Non-digital activities could be objects that could not be embedded in the digital building model in the form of extra work not included in the original agreement or building objects that provided a service.

Since the smart contracts could be connected to the blockchain, the smart contracts became fundamental. Therefore, the smart contract was regulation, based on fixed routines and coding processes, which governed the blockchain technology. The execution of the smart contract took place through the operations that were linked to the respective blockchains. There was the understanding that the smart contracts differed from a traditional contract.

The smart contract acted as a digital representation of the normal contract, although with the challenge that the smart contracts were more rigid than the written contract. The smart contract did not have the same flexibility as the written contract, as the traditional contract was based on written language that could be ambiguously interpreted. Construction blockchain was governed by the community logic with a focus on the trust creation that was in blockchain and therefore also in the trust that was in the smart contracts and the trust that it would be executed correctly and according to the agreed schedule. As described in the analysis of both build trust and BIM partner, the smart contract was governed by regulatory legitimacy, through algorithmic laws and rules. In time, the smart contract could be pragmatically legitimised when the construction industry discovers the benefits of a trust-building contract.

In order for the construction industry to make use of BIM partner and build trust, the construction blockchain must clarify for the industry, that the project could create the necessary changes in relation to digitalisation. They could do that to show that build trust and BIM Partner were value-creating, trust-based and supportive of the use of smart contract. It was important that emphasis was placed on the tender material being clearly and unambiguously worded and that no changes could be made in relation to the basis. Clearly and unambiguously formulated was one described in the standard contract, which has been the basis for the client losing the dispute.

Once this was done, mobilisation of the other parties in the industry could be discussed. It was important that construction's blockchain described their work as innovatively developing and helping to change the way in which we perceived digitisation of the entire construction value-chain, including the construction contracts, and so mobilising the innovative parties in the construction industry.

The Construction Blockchain

So, we can ask can the contract's legitimacy problem be remedied with smart contracts. The construction blockchain believes that the problem with construction contracts could be remedied with technology. Through their pioneering projects, construction blockchain had shown that blockchain was based on trust in non-editable data, through which smart contract could work. Through the use of blockchain technology in build trust, the construction parties could get a more precise and unambiguous tender material, without the risk of using or giving an offer on a non-validated tender material, where the tender documents help create trust across the organisations.

Build trust is therefore based on pragmatic legitimacy, where it is recognition of the organisations and benefit from each other and each other's legitimacy through trust in the comprehensiveness of the tender materials. The problem with unfinished and non-buildable digital building models would be solved by the Building Authority's Blockchains through the pioneering project BIM partner. The building's blockchain will use BIM partner and, through the platform, verify the digital building model in relation to the client's requirements and in accordance with the consultant agreement entered into. Therefore, clients could have their requirements met, without having to check each and every drawing, by reading the report of the data extraction that was carried out.

Clients also got a digital building model that subsequently could form a basis in relation to the desired priority in the construction contract and could so be used throughout the construction process. The construction blockchain was therefore helping to set the framework for the future use of blockchain in the construction industry and through the network of participants in the construction industry. Blockchain had proven to be useful in relation to the tender material and validation of the digital building model. As Build Trust, BIM Partner was driven by pragmatic legitimacy.

Construction Blockchain and Smart Contract

A statement from HD-lab Head of Consulting & Innovation, and developer of BIM partner, Niels Falk, was the next step in the use of blockchain as an implementation of smart contracts. Since build trust and BIM partner used blockchain as a technological basis, a smart contracts could be linked on both platforms. At build trust a smart contract could give the tendering contractor a quick overview of the tender documents. By connecting the tender documents to the smart contracts, the client was forced to deal with whether the tender material complies with requirements for clear and unambiguous wording.

When the tender documents were connected to a smart contract, these could not be changed and so the contractor offered the service described. This was a help increase confidence that the tender material had been validated and checked. The builder simply put it into a situation that everything that was not coded with blockchain could be a risk performed as extra work. BIM partner was used for validating the digital building model through blockchain and could thereby be connected to a smart contract. When blockchain was linked to the digital building model, this could be used as a basis for the tender material. This could help create confidence that the digital building model had been executed according to the client's requirements and thereby became part of the tender material and formed the basis for future execution.

In order for the digital building model to be used for all construction contracts, it was necessary that all building parts were included. Smart contract could not handle non-digital activities, therefore non-digital activities would be perceived as extra work on the part of the contractor, as they were not included in the contract basis. Construction blockchain's pioneering projects were legitimised through the trust inherent in the non-editable data. The contractor bid on what was included in the smart contract and could use the digital building model to validate materials and performance through the building objects coded with the correct blockchain.

CHAPTER 19

Digital Twins

As-built sounds like a throw-away comment, but it is nothing like that. It is an exact virtual copy of a real physical entity. It can be collated in an *ad hoc* method or it can be part and parcel of the deliverables. Either way it is an expensive exercise, because of the differences that can occur after tendering, where LOD350, 400 and 500 (as-built) can frequently change and not be properly documented.

But digital methods are making this transition easier. Some building sites scan the site regularly and overlay the result on to the 3D model. A simple exercise can clash detect the two entities and areas of clashes can be identified and the correct solution resolved. Whether this happens automatically or is addressed manually is also changing. But the benefits are plentiful, having a virtual copy of a project allows many other take-offs to be conducted, bringing us into facilities management and life cycle analysis to be verified and tested.

Depending on how the building is administered, many answers can be given for actual settings and what-if scenarios. If the building is sensored and energy usage monitored, then careful control over costs

Transforming the Construction Industry with Blockchain: Enhancing Efficiency, Transparency, and Collaboration, First Edition. James Harty.

can be made, as it is the running of a building through its life is where the biggest costs occur. The Edge in Amsterdam is a very good example of this in practice. By accumulating data from the building, a matrix dashboard can control who is where and when, and consequently plan the space and day-to-day occupancy better.

Traditionally, facilities management happened after handover, and from scratch, precisely because of the changes that happened through the construction work. A clean-sheet approach was seen as a fresh start and the delivered entities were all that was left, suggesting a shorter and focused approach. Each room in the facility would be visited and each entity would be recorded on a clip board, while noting how many light fittings there were. I know I am being glib here, but there was no flow in their procedures, and it represented a lot of double-work.

The financial model is a process usually to find a decision to build or not, it involves a financial institution or at least a financial plan. What is it going to cost, what is the budget and how is it going to be paid all come into the mix. To appraise these issues, a notional building is addressed where occupancy, function, location and their impact is assessed on a spread sheet, where the building's form is not part of the equation, at least not until the money is approved. The people making these decisions are usually not spatial or graphical in their prowess and any hint of form is unwelcome and ill-advised.

But adjacency and interaction can be an important part of this process and this is often represented through a bubble diagram. Large bubbles represent large spaces and often are accompanied by notional areas or numbers of occupancy, and these can be overlapping or connected by lines. Placing a massing element on the site with the desired height, or placing a parametric volume, which maintains the square metres floor area, room separators can be used to generate circles (essentially free forms), rather than using the walls as room delimitators.

These can then be named, tagged and sized, from which schedules can be drawn-off, but critically still contained within the model, cutting out double work. This process happens without defining rectangular areas, which can often be misread as definitive spaces so that the abstract nature of the forms can be maintained. When the correct mix is found, price books can be associated with the data and budgetary figures are determined. Schedules can be made in ways better understood to financial people.

This work is phased within the model as preconstruction work. This means that it can co-exist within the model proper when construction work is subsequently prepared. The benefit of having it here is that specific climatic data can be added to this conceptual form and feedback given, regarding shape, orientation, shading, heat gain, exposure and energy performance. A report can be generated containing all the above data and if several forms are tested, several reports can be generated and cross-referenced in a compare and contrast fashion, giving informed comment.

Next, with the design model, we are armed with this data and the next phase of procurement is now well prepared. As the design progresses the early work is not lost and as each form becomes an entity, the early data is kept and updated, and reports can affirm compliance with the initially agreed proposal. For the client, this gives a greater amount of certainty to the project, which can be lost in traditional procurement methods. Scanning the building site during construction allows the model and building to be synchronised with each other.

If the above has been carried out as described, then the final virtual model will be a replica of the actual building and become the facilities management model. Within the model, each element has a right click properties dialogue box built up of parameters and values. If the contractor uses the model to enter the supplier, then other data can be added such as durability, colour, model number as well as all ironmongery and key identification numbering. Or if it is not done, then that the ability is there for later or whenever appropriate. This is the beauty of a centralised point of contact.

Just as smart phones use apps (applications) to do things, bots (robots) are waiting to do other things. Search engines are very well-advanced today. Enter a word or topic in your browser search engine and a meaningful response is returned based on others who made the same search and relevant to your location. All this happens in the background and without going into the algorithms, we all use it and are relatively pleased with the hit rate and response.

All-in-all there are fascinating developments happening, and they are happening at a rate of knots. Patrick MacLeamy has engaged us with his 'BIM, BAM, BOOM' scenario where for every dollar spent in the design phase, there are $20 spent in the assembly or construction, which leads to $60 in its operation and maintenance. If clients and users are not demanding this consideration in their projects, then we are failing them. If we are not looking at sustainable issues through all phases, then we are falling ourselves. As said before, incentivising and rewarding better practices leads to better built environments and better quality in our building stock.

Robotics (Scanning and Clash Detection), The Mechanics of How It Will Be Implemented

Retail fashion and cosmetics are going through a minor revolution these days with magic mirrors. This is a form of augmented reality, where the person before the mirror can see their image augmented to promote a product. Typically, a person picking up a lipstick, magically sees the precise hue of colour on their lips. Changing from crimson red to frigid pink happens before their very eyes. Picking a silk blouse from the cloths rail, magically transposes the blouse on to the person in their size, one click away from a purchase.

Taking a child into a Lego store is no-less also undergoing a seismic change. Children have no problem taking packages off the shelf, holding them under the camera/sensor above the monitor, only to see the contents of the box magically built before their very eyes in a mixed reality experience. Parents look on aghast, while the kids get it straight away.

Both these virtual realities are slowly making their way into construction, and about time too. Already products can be transposed into settings, such as an armchair placed in a room while the various colours and features can be evaluated before buying. IKEA too is moving its assembly manuals to an augmented reality app, which locates the flat pack in the room and step by step shows how to assemble the chest of drawers or whatever, paying particular attention to difficult or not obvious details (Warren et al. 2018).

Construction smart hard-hats (with a visor), on to which project information can be displayed (via a dashboard) or augmented virtual modelling can be placed into their position in real time are now available (Teicholz 2004). On a lessor level, handheld devices can perform the same tasks and either by using QR codes or barcodes can retrieve relevant information to relevant tasks at the source of the problem (Pauli 2022).

CHAPTER 20

Conclusion

Back in the early 1980s, I wrote a six-page report on a typewriter. It was my friend's sister's and up until then I had written everything in long hand. I had good handwriting, today if I do not enter my scribbles into a digital device, I cannot read what I have written the following day. Using the typewriter there were many, and I mean many restarts, removing sheets of paper and consigning them to the bin, and even when something had been procured, there was a need to reread and check for spelling mistakes. These could be corrected using white-over or simply a biro in the margin.

In the late 1980s, I wrote my master's dissertation on a computer. Essentially, the same options presented themselves, except that corrections could be made on the screen. But it was still me proofing and accepting, being the person of probate. Formatting allowed double column pages with headers, footers, footnotes, references and citations. There was even spell checking, but it was passive and required your personal intervention and approval.

Transforming the Construction Industry with Blockchain: Enhancing Efficiency, Transparency, and Collaboration, First Edition. James Harty.

My next milestone was my PhD, in the noughties, predictive text was available. We never questioned or reacted to this. It knew whether it was British English, American English or Esperanto. But at each and every stage along this process, we are relinquishing our authority and letting the machine take over. Navigation became easier, using text styles (headings) to control formatting. The timeline might not be totally accurate but the process tells a story, I want to convey.

Fast forward to me writing this book, midway through the activity, artificial intelligence and machine learning hit the proverbial fan. Suddenly, ChatGPT could generate works that used generative pretrained transformers (GPTs). These are large language models that use semantic relationships to predict works, and they are unleashing themselves on the world, even with the horrendous hallucinations (falsehoods) that occur as it builds its quality data.

I can tell AI who I am, what I do and I can upload all my publications, and from this data, it can produce work akin to my position and role. This leads to issues of copyright and fake news. I can also upload my current work and ask for a review, stressing it be critical, positive or any slant I care to apply. At time of going to press, The New York Times is challenging ChatGPT, about how it has acquired it archival data from behind a paywall, and whether it has the right to do and promote this?

This has happened to the written word, through my academic life. This same process will now descend on construction and for better or worse have a profound effect on it. In order to maintain verification and validity, blockchain will enter the fray as the single source of truth (SSoT). It can trump AI and ML, but as their model grows and improves, it will regulate the process and an alignment should emerge of all these new technologies.

This is a good example of the process that we currently find ourselves in. Curbing its provenance is pointless, we must own it as a skill. It is no worse than a search engine on steroids. Learning to work with it positively is a must, and as I said earlier 'AI will not take your job, people using it will' (Doherty 2023).

Disruptive Technologies, How It Will All Settle

John Tobin, in his article: 'Measuring BIM's Disruption: Understanding Value Networks of BIM/VDC' stated:

> *'BIM is a disruptive technology, in contrast to CAD, and brings a different 'value network' to the AEC industry. . . . where BIM sows*

information in a model environment, while VDC largely harvests that information for downstream uses including commissioning, facility management and construction logistics' (Tobin 2013).

He refers to CAD being a progressive feature of traditional methods, that it was merely computer-aided drafting, whereas VDC harvests data for several new features, not fully understood by the profession at that time. David Shepherd, author of the BIM Management Handbook, in an article entitled 'Ahead of the Game' (Malleson 2016) stated:

> *'Where a disruptive technology emerges – and BIM is a disruptive technology – it's effects on the mainstream is not always clear. What we might be seeing is the early stages of disruption, and in those early stages it's very difficult to know what the effects will be. We don't have the breadth of vision to see where, and for whom, the benefits of BIM will emerge'* (Malleson 2016).

The most notorious example of new technologies is in the car industry. Before the advent of automobiles, the equine (horse) trade had a plethora of support industries supplying and serving the splendid beast. Saddlery was a bespoke industry supplying handcrafted saddles to all and sundry, covering both gentry and pauper who both needed a horse. There were bridleries and blacksmiths, all in service, and ultimately hitching posts for stagecoaches to reload horsepower and replenish travel-weary passengers.

All changed with the advent of the all-conquering automobile, the leather workers moved on to factory assemblies, producing leather seating and trims. Blacksmiths diversified into metal works. Hitching posts became petrol stations and roadside eateries. The ultimate insult was Bugatti who reduced the symbol of the horseshoe influence, into the shape of its radiator grill badge.

More recently, in the newspaper industry, the 80s also experienced a most cruel come-uppance. Typesetters, basically, became redundant, as computers replaced them in formatting newsprint. The Wapping dispute in 1986 saw Rupert Murdock take on the might of the unions, as he implemented a clandestine manoeuvre, building a huge printing plant, unbeknownst to them, in Wapping, east London (Neil 2006). Indeed, not only typesetters but also compositors, linotype operators, machine room personnel, publishing room employees, clerical staff and copytakers numbered the 5,500 workers who were called out on strike by the then very strong print unions, who like the coalminers a year before ultimately perished.

The assimilation of the journalists away from the 150-year-old 'hot metal' system to the new streamlined digital method was also interesting. Tony Bevans, The Times political correspondent told his colleagues, totally deadpan:

> *'I want my stories to be published . . . When you are faced with overwhelming pressure, and you are in a cul-de-sac . . . you must either fight to the death or lie down . . . we have a gun to our heads. I believe most of you will go with ashes in your mouths'* (Macintyre 2016).

Such a scenario could be framed for the current situation in the construction sector, of those seeing the media of choice as paper, with all its support structures and the digital solution, with which they do not want to engage (into meaningful discussion), and those reaping the benefits of collaboration and trust, bringing certainty and surety to the engagement (Harty and Laing 2011). A postscript to this is that the printed newspaper today is seeing a huge demise, with many failing to attract sufficient advertising, and few readers being willing to pay for a subscription.

Project Work

If we can say that one of the primary objectives of vocational educations follows directly from the nature of professional work it purports to serve, then project work moves discovery and application to centre stage. It becomes the application of knowledge to ill-posed problems (Thacher and Compeau 1999). Interdisciplinary and problem-based learning often rile against traditional academic learning, whereas somewhat ironically, vocational educations seem to embrace them wholesomely (Olesen and Jensen 1999). The impact of machine learning will challenge this point but I cannot comment on how.

Such is the case in Denmark, where Roskilde University (RUC) is one of the most radical reform universities in Europe (Olesen and Jensen 1999). In the 70s, a number of new universities promoted alternative concepts of teaching (Innovative Student-Centred Education) where radical and political reforms brought new teaching techniques. These methods have found their way into the Copenhagen School of Design and Technology (KEA 2013; Øhrstrøm et al. 2013).

Project work endeavours to bring a modern profile to independent analysis, problem-solving and training in co-operation, involving; complex issues, critical attitudes, political awareness, responsibility, professional commitment and last but not least, overview

(Illeris 1999). Learning and problem-solving do not happen in a vacuum (Wildemeersch 1991), meaning that they need a context.

Learning by doing provides this context, which is the great dictate from the Chicago School (Dewey 1916). Teaching and learning have, in common, an ambition to combine elements of problem-solving and learning in a dialectic way. If traditional learning desires that the student learns everything about a given subject, leading to subject matter abundance, then project work desires exemplarity, in learning by doing. While project work can be the vehicle for teaching skills and knowledge, it also provides a very important latent means of developing general professional skills and attitudes (Kerszman 1999).

Two types of experiential learning can be identified in 'learning by doing', the first is 'direct encounter with the phenomena being studied rather than merely thinking about the encounter, or only considering the possibility of doing something about it'. (Brookfield 1983). This occurs mainly in professional courses. The other 'occurs as a direct participation in the events of life'. This is achieved by reflecting on our experiences and is typical of learning in general.

All Change

Our education will transform drastically in the next 5 years. In the next 2–3 years, there will be a pedagogic transition where our students will use the new tools, namely artificial intelligence and machine learning, to make all deliverables. There is merit in this hybrid approach and it will be positive. But in 5 years' time, machine learning will provide everything without the need for human intervention. This means that when given an assignment, uploading to AI will produce all aspects of the deliverables, meaning that the student can spend the semester on the beach, showing up for the exam only.

We are seeing the loss of many jobs due to this new paradigm, and this is happening in law, medicine and many applied hands-on careers. There needs to be an olive and branch approach where the essence of the education is appraised and nurtured. What is critical here is if we do not start now, we are lost. Within construction, all areas of teaching require deliverables which are the product of knowledge, skills and competences.

In BIM, large language models will begin to automate tagging and dimensioning, number rooms, spaces and elements, name and standardise drawings, create code (python/dynamo), map and organise parameters, assist in the delivery scope, perform quantity take-offs, automate view creation while most importantly automate deliverables.

It can create BEP templates, and implement them, applying processes, mapping activities and tasks, distil standards as they are developed, understand the process and aid its implementation, support scheduling, create documents, ensure compliance, extract and translate documentation.

It can also validate data exports, analysis performance and report both good and poor periods, sort and visualise data, search quickly for information, create repositories, research all documentation, assure quality and control, organise and name all files. In educational terms, it can teach standards, form workshops, brainstorm, summarise events and navigate family/element creation. Finally, it can be of general assistance in text, imaging and/or fly-throughs, take notes and action tasks, recommend tools and relevant software, automate workflows and compare and contrast documented versions and revisions.

In design, architectural deliverables are all drawings (plans, sections and elevations) as well as renderings and fly-throughs. For process and planning, all documents will be automatically generated. In structural engineering, the application of load and sizing will be automated, while in services, all ducting and sizing will also be automated. Materials offers a slim chance of survival because of the innovative aspect of the subject, but much of the selection process will be automated. As machine learning improves the process becomes more streamlined and efficient.

Otherwise, all stakeholders will seamlessly interact through *data only*. The design team will dovetail with the municipality for all planning issues and the tender will be awarded through prequalification and a certainty of the quality required and delivered. Financial institutions will appropriate costings and reap the benefit of a sharper, better honed product. The role of robotics has not even been mentioned in this process. But ultimately, this has immense consequences for how we, as humans, will entertain ourselves or retrain our purpose.

Already, we are seeing the proliferation of ChatGPT in much of the student deliverables, and owning this new technology is important for the future. If we do not attend to this new position, then the education will reduce to, at best, a 2-year bachelor's degree. In the bigger picture, financial institutions will see no purpose in soliciting all workstages and paying for each handover, when they can go direct to the digital twin, and save money. Once stakeholders are named or sought, AI will name check and prequalify them for their ability to perform. The type of contract, the limitations all will be assessed with commutable decisions, done with robustness and verifiable outcomes.

I do not wish to be the harbinger of doom, sticking our heads in the sand is not the solution either. We need to engage with industry and plot a way forward. Talking to industry, it is acutely aware, so there is a sea of goodwill, in which we can engage. This is positive and understood. But what is patently clear is that we are looking at a sea of change and great developments in the not-too-distant future.

One area that offers great potential is innovation and the adoption of recycling/upcycling of what is regarded today as waste. Currently, there is no proper methods for assessing waste and proving the robustness of the proposed solutions. This will require much research by the trying and testing of various proposals, to consider their worth. Often, this requires significant outlays of finance to establish validity and this needs to be nurtured and given the space it needs to blossom and find new methods to reuse materials. This might also impact processes and other aspects of construction today.

Innovation

But all is not lost, Anders Lendager is an architect in Copenhagen who devotes much of his time to sustainable solutions where he upcycles/recycles materials and reuses construction waste to optimise the material palette. While sounding straightforward, it is a method fraught with obstacles, mostly verifying the fitness-for-purpose of the reused material. This often means a lead-in loss maker to establish the credentials of the material. This means bringing a new mindset to construction in an extreme situation where he goes above and beyond what is considered a valid material.

Anders Morgenthaler and Anders Lendager have a VLOG, 'Reconstructing The World' where they post inspiring videos on how to reconstruct the world through more thoughtful architecture. One such video looks at lava as a new sustainable material (Lendager and Pálmadóttir 2023). Anders Lendager and Arnhildur Pálmadóttir stand in a lava landscape in Iceland and discuss the material as a natural alternative to concrete. First there is a discussion of what lava is, where we are told it is a molten rock (magma) that is expelled from the interior of earth on to its surface.

When a volcano explodes, it is seen as dangerous and usually causes much damage on to the landscape that it covers. It is a basalt rock with a mix of aluminium, calcium, magnesium, iron sodium and potassium with a mixture of trace elements. If it comes through water, it can be like a pumice stone which is light and a good insulator. If it comes directly on to the landscape it can be a hard heavy basalt, when it cools, and this is like a concrete material, offered free by mother earth.

So, if it is properly harvested it can offer four different products. First as a pumice stone, it can offer an insulating material, this can be carved-out, planed, crushed and be used in a floor, roof or wall. Second, using the harder stone, it could cut out as columns and beams as building components from the lava flow field. So, this potentially can form all the building components needed for the construction. Third, it is to dig trenches in the earth and when the eruption occurs, there are rivers of molten lava, which are massaged into these channels like in situ-concrete. Once these channels are cooled, the earth can be removed, revealing structural walls.

This has been used traditionally in an old village where the lava was used to create walls to make cottages quite successfully. Lastly, there is a sweet spot under the surface of the cooling lava, which could be tapped and piped to construction sites where it can be used as a type of 3D printing. It can also be poured into moulds to make building elements.

So, the major costs are cutting the material, sawing through it and this uses locally produced electricity, which is cheap, because of geothermal qualities in Iceland. Climate Watch notes the total emissions of the country is 3.27 $MtCO_2e$. To put this in context USA is 5,289.13 $MtCO_2e$ and China is 12,295.62 $MtCO_2e$ (World Resources Institute 2024). India is 3,166.95 $MtCO_2e$, which has been equated to how much cement (concrete) releases if it was a country (the third worst emitter).

Arnhildur comments that mother earth is giving us this for free, asking why we do not use it. Transported locally it is very carbon dioxide friendly. But it might be also exportable, which is ironic because, in Iceland, most building materials are imported. Another aspect is that it could be used anywhere that there are volcanos (Hawaii, Japan, Scilly . . .).

In another scenario, they have managed to demonstrate that a 20-storey building can be made of timber. It is currently on site where the greatest hurdle was showing a robust fire strategy. Currently, the tallest building (allowed in Denmark) in timber is three storey, precisely because of fire safety. They spent over a year and a half researching how to address these issue. Their research showed that timber chars at a rate of 1 millimeter per minute when exposed to fire. Therefore, to get 30 minutes fire resistance, an extra 30 mm's are added to all sides of the beam, column or building component to comply, and 60 would be 60 mm's. This means that the columns used in the building are 400 × 400 mm's, made of glulam timber.

They could have provided core lift shafts and fire escape stairs at a little over 800 mm's, but the client felt that it was costing too much in net floor area, so the cores are in concrete, meaning that it is a hybrid building. There are 4,000 m^3 of timber in the building, 1,000 of which is waste wood. This means that they have 1.5 kg/m^2 for the whole building. To reach 70% reduction as required in the Paris agreement, would mean 2.5 kg/m^2 meaning it is way below the target.

Problem-Based Learning

Some see problem-based learning as a pedagogical means of ensuring continuity and focus for the learner. That is, instead of being presented with random pieces of information, it points out a definite direction for the student using the project as a vehicle. Ultimately, in societal terms, it educates critical technocrats and problem solvers who are not afraid to defy disciplinary limits, precisely because real problems do not respect disciplinary limits (Flodström 1999).

As construction projects become more complex, solving them requires more interaction between many people. This too necessitates certain skills and personal characteristics, making the participants more creative and independent (ibid). Indeed, the uncertainty, complexity, instability, uniqueness and value conflict that occurs during a project, questions the very goals themselves (Schön 1987).

One of the best reasons for group work is the possibility of exercising the student in the vagaries of collaborating in software such as Revit, notably in using Worksets and Central files. It can be very difficult for other institutions to initiate such an arrangement, without strenuous and complicated set-ups, but here, the set-up welcomes it. The positive response here was a little surprising but reflects well on the classes' motivation. Primarily, it increases co-ordination and helps communication enormously (Harty 2012). The educational licensing system means that each student has access to the whole suite of programmes on offer from Autodesk, ranging from AutoCAD light right through to Inventor Fusion and all in-between, offering student licenses per programme (Autodesk 2015).

The school's course syllabus is making in-roads in this direction, looking to introduce an appreciation of accountable sustainable mechanisms from the first semester to parallel elective classes in Facilities Management in the fourth semester. This aspect is further being tackled at the school, trying to harness the demands and requirements of the FM process within the BIM process. This part is most interesting with regard to the education that is offered and their job opportunities upon graduation.

Group Work

Further to project work and problem-based learning is the notion of group-work. A common form of group work is the beehive, where students having different skills have to co-operate to solve tasks (Chiriac 1999). Steiner defines three conditions for group work: task demands, resources of the group and the process, by which the group performs in order to complete the task (Steiner 1972). Since the adoption of BIM over CAD in the classroom, the stress levels of the students, in group work, have significantly diminished because all can easily see the model and see what is required or fix what is wrong and does not work.

Group work sounds very appropriate, it mirrors the work place and involves collaboration. But when a whole semester is based on the same group, with inter-dependency for grades and marking, the stakes are high, and this reflects in how very much more seriously, the students take the matter. While putting them in a very strong position to collaborate, it also gives them a better understanding of integrated project delivery (IPD).

Furthermore, using 'analysis' software, these scenarios can be tested, and reports generated showing compliance to regulations or itemising nonconforming failures. Whether it is sustainability, economics, building codes or local plans, certainty and control are in the hands of the designer and the project team. Under sustainability, the building's performance can be modelled with regard to energy use, passive heating and cooling, together with form, orientation and layout. Economy is quite simple, extract quantities and price them (Harty and Laing 2010). BuildingSmart has already demonstrated that building codes can be assimilated into the approval process (Nesbitt 2006).

How Would You Get a Notorious Non-payer To Step Up to the Plate?

Construction is a cut-throat business and often if someone can pull a fast one on another, there is no love lost. This is manifest in the non-payment and delayed payment across the sector. It can also be short payment or reduced sums being paid. So, for blockchain to regulate this area of concern there is a demand, requirement that funds and financing have been acquired, and that it be placed in the hands of the app to appropriate and sign off work as and when complete.

Releasing control of funds can be an anathema for many players and a downright no-go area for some. The benefit for complying with this is a huge mindset change. So, how can this be best sold without having prior experience and affirmation of its worth? Rarely, is finance already available and usually it has to be negotiated through banks or similar institutions and this is where the pressure can emanate. Banks need to see the benefit here and push for its implementation. Public procurement can also play a part here, requiring it as a procurement method.

To repeat, Eurotunnel had difficulties in motivating the suppliers once the contract had been awarded called 'moral hazard' (Winch 2002). Essentially, the client is somewhat unsure that the contractor will fully mobilise its capabilities on the client's behalf, rather than its own interests or some other third party. The preferred option he calls 'consummate performance' instead of more likely 'perfunctory performance'. Here two cultures collide, with the banks preferring to move the contractor to a fixed price, While the contractor works on the basis that the estimates have to be low, to ensure that the project gets commissioned.

It can be addressed as part of the contract method and written into the agreement. The banks can demand that the contract sum be placed into an account where it can be administered. Of course, there has to be safety precautions and a possibility to appeal, but not in the litigious procedures currently in practice. As said previously, there has to be a no-blame culture and 'a find it as it is' mentality to promote these methods. This can be seen on SISK's methods (Center Parcs) and John Egan's Rethinking Construction (Heathrow Terminal 5).

Ironically, Center Parcs subsequently adopted this new procurement method across its whole portfolio, while Terminal 5 resorted to other methods in their satellite buildings (Terminals 5A and 5B). Adopting these new methods will also promote better practices and will bring life cycle assessments and sustainability into the picture. Blockchain can also foster incentivisation and reward to bring added benefits into play. Providing the building stock, we need rather what is the least return that needs to be made, improves baselines and offers better profit margins.

Machine learning will also improve the method of implementation as it learns from procedures in similar circumstances. This will also free-up time for design work and aesthetics. Performance takes centre stage too, offering and demonstrating better execution. The potential and gains-made will catapult the industry into the next phase of construction, improving all aspects of how we build today. It is exciting and a revolution in the making, enjoy.

As said at the outset, construction professionals have often strived to recover the intrinsic value of their labour. Blockchain offers a new value proposition to extract reward not just for the collaborative services but also the intrinsic intangible value across the life cycle of a facility. It becomes the gift that keeps giving, a contract that rewards value, a contract that does not reward non-performance (which is just as important) and an agreement that releases payment when due or expected. Together with AI and ML, there is a splendid future to behold.

References

AIBINU, A.A., CARTER, S., FRANCIS, V. and SERRA, P.V., 2018. Necessary evils: Controlling Requests for Information (RFIs) to reduce cost and improve margins. *Construction Research and Innovation*, **9**(4), pp. 103–108.

ALLEGRETTI, A., 2021. David Cameron said to have made about $10m from Greensill capital. *The Guardian* (09-08-2021).

AUTODESK, 2015-last update, **Autodesk** education community. Free software [Homepage of Autodesk], [Online]. Available: http://www.autodesk.com/education/free-software/all [10-07-2015].

BARRATT, L., 21 June 2018-last update, Grenfell council was warned about gaps in windows by resident of flat where fire began [Homepage of Inside Housing], [Online]. Available: https://www.insidehousing.co.uk/news/news/grenfell-council-was-warned-about-gaps-in-windows-by-resident-of-flat-where-fire-began-56882 [13-06-2019].

BARRETT, P., 2008. *Revaluing Construction*. First edn. Oxford: Blackwell Publishing Ltd.

BARRETT, N., 2010. ***The Role of the Architectural Technologist:*** *An Assessment of its History, Traditions & Impact on The Modern Construction Team*, Hovedland, Denmark.

BASU, A. and JARNAGIN, C., 2008. How to tap IT's hidden potential. *Wall Street Journal*, **2011** (25-07-2011).

BECK, R., KUBACH, M., JØRGENSEN, K.P., SELLUNG, R., SCHUNK, C. and GENTILE, L., 2019. *Study on the Economic Impact of Blockchain on the Danish Industry and Labour Market*. TR-2019-206. Copenhagen: IT University of Copenhagen.

BENNETTS, R., 2010. Divide and fall. *RIBA Journal*, **2010**(10): 22.

BESWICK, P. and BAILEY, L. 2019. *The Global Risks Report 2019*. 14th Edition. Geneva: World Economic Forum.

BEW, M. and UNDERWOOD, J., 2010. Delivering BIM to the UK market. In: J. UNDERWOOD and U. ISIKDAG, eds, ***Handbook of Research on Building Information Modeling and Construction Informatics:*** *Concepts and Technologies*. Hershey, PA: Information Science Reference (an imprint of IGI Global), pp. 30–64.

BIM-FORUM, 2017. ***Level of Development*** *Specification Guide*. Dallas, TX: IKERD. November 2017.

Transforming the Construction Industry with Blockchain: Enhancing Efficiency, Transparency, and Collaboration, First Edition. James Harty.

BOOTH, R., 14 January 2022-last update, 'A merry-go-round of buck-passing': Inside the four-year Grenfell inquiry [Homepage of The Cuardian], [Online]. Available: https://www.theguardian.com/uk-news/2022/jun/14/a-merry-go-round-of-buck-passing-inside-the-four-year-grenfell-inquiry [04-04-2023].

BRAHNEY, A., 28 October 2019-last update, Workshops: SMART UX: Engineering user experience. Available: https://2019.ctbuh.org/presentations/?paper_id=345 [04-11-2023].

BREEAM, 12 June 2019-last update, **BREEAM** Category issues & aims [Homepage of BREEAM], [Online]. Available: https://www.breeam.com/discover/how-breeam-certification-works/ [12-06-2019].

BROOKFIELD, S.D., 1983. *Adult Learning, Adult Education and the Community*. Berkshire: Open University Press.

BROOKLYN ENERGY, 2019-last update, **Brooklyn** *Micro-grid* [Homepage of Brooklyn. Energy], [Online]. Available: https://www.brooklyn.energy/ [12-06-2019].

BROSHAR, M., STRONG, N. and FRIEDMAN, D.S., 2006. *Report on Integrated Practice*. Washington, DC: AIA.

BROWN, S., 2019-last update, What can insurers do to improve the way they assess and manage the short and longer-term impacts of climate change? [Homepage of KPMG], [Online]. Available: https://home.kpmg/xx/en/home/insights/2019/03/combating-climate-risks-the-future-of-insurance-fs.html [28 Dec, 2020].

BROWN, K.A., 2023. ***The Decline of Sustainability Values Throughout the Development of AEC Building Projects.*** *How and Why Are the Sustainability Goals That Are Set at the Inception of a Project Lost Throughout the Process?* Copenhagen: Copenhagen School of Design & Technology.

BSI, March 2014-last update, **PAS 1192-2:2013** Specification for information management for the capital/delivery phase of construction projects using building information modelling [Homepage of BSI], [Online]. Available: http://shop.bsigroup.com/Navigate-by/PAS/PAS-1192-22013/ [03-09-2014].

BSRIA, 2014-last update, **The soft-landings** – Core principles [Homepage of BSRIA], [Online]. Available: www.bsria.co.uk [04-03-2014].

BUFFA, M. and EASTMAN, C., 2014. **Building information modeling** – Case study (Fondation Louis Vuitton). In: ARCHITECTURE DEPARTMENT SOUTHERN POLYTECHNIC STATE UNIVERSITY GEORGIA, ed., *Session C1. Digital Exploration: Visualization, Education, BIM 2014*. Bozeman, MT: Design Communication Association, pp. 125–133.

CALA-OR, C., 2023. *Responsive Smart Building Built Environment and Data Collection*. KEA.

CALVERT, N., 2013-last update, Why we care about BIM . . . [Homepage of Directions Magazine], [Online]. Available: http://www.directionsmag.com/entry/why-we-care-about-bim/368436 [03-28-2015].

CARROON, J., 2010. ***Sustainable Preservation:*** *Greening Existing Buildings*. Hoboken, NJ: Wiley.

CFMA, 2011-last update, What is typical profit margin general contractor? [Homepage of Construction Financial Management Association], [Online]. Available: www.cfma.org [16-03-2020].

CHIRIAC, E.H., 1999. Group-work in traditional higher education. In: H.S. OLESEN and J.H. JENSEN, eds, ***Project Studies*** *– A Late Modern University Reform?* Frederiksberg: Roskilde University Press, pp. 293–303.

CICMIL, S. and MARSHALL, D., 2005. **Insights into collaboration at the project level**: Complexity, social interaction and procurement mechanisms. *Building Research and Information*, **33**(6), pp. 523–535.

COFFEE, T., October 2006-last update, **The future of integrated concurrent engineering in spacecraft design**. The lean aerospace initiative working paper series [Homepage of MIT], [Online]. Available: http://web.mit.edu/tcoffee/www/docs/lai-cet-rsp-ICEToolsStudy-tcoffee-8a.pdf [January 2016].

CONFESSORE, N., 2017. **Too rich for conflicts?** Trump appointees may have many, seen and unseen. *New York Times*, **A15**, 9.

CONSOLIS, 2023-last update, **BLOX**: A major development project in Copenhagen [Homepage of Consolis], [Online]. Available: https://www.consolis.com/blox-a-major-development-project-in-copenhagen/ [01-05-2023].

CONSTRUCTION SITE, 2010-last update, Contract procurement methods [Homepage of Construction site], [Online]. Available: www.constructionsite.org.uk/repository/resource/view_resource.php?id=10 [07-10-2010].

CROWN, 2009-last update, Office of Government Commerce [Homepage of © Crown Copyright], [Online]. Available: http://www.ogc.gov.uk/index.asp [09-08-2011].

CUMMING, D., 2020-last update, Why asset managers cannot be passive on climate change [Homepage of AVIVA Investors], [Online]. Available: https://www.avivainvestors.com/en-gb/views/aiq-investment-thinking/2020/02/climate-change-and-ceos/ [08-04-2020].

DESIGNING BUILDINGS WIKI, 28 March, 2019-last update, Information and communications technology in construction [Homepage of Designing Buildings Wiki], [Online]. Available: https://www.designingbuildings.co.uk/wiki/Information_and_communications_technology_in_construction [10-06-2019].

DEUTSCH, R., 2011-last update, Beyond BIM Boundaries [Homepage of Design Intelligence], [Online]. Available: http://www.di.net/articles/archive/bim_beyond_boundaries/ [2nd August, 2011].

DEVLIN, K., 2021. Tory donor drawn into lobbying scandal after being handed key business role. *The Independent*, **UK Politics**.

DEWEY, J., 1916. ***Democracy and Education***. *An Introduction to the Philosophy of Education*. New York: Macmillan.

DOHERTY, P., 2023. Unlocking Thee Metaverse: Digital Real Estate, BIM and the AEC Industry. *BIM Coordinators Summit* [07-09-2023].

DOUNAS, T. and LOMBARDI, D., eds, 2022. *Blockchain for Construction*. Singapore: Springer.

Dunne, D. 2020, 28 Dec. Ten worst extreme weather events in 2020 cost world about $140bn, report says. *The Independent* https://www.independent.co.uk/environment/climate-change/extreme-weather-events-2020-climate-b1778158.html.

EASTMAN, C., TEICHOLZ, P., SACKS, R. and LISTON, K., 2008. ***BIM Handbook*** *A Guide to Building Information Modeling for Owners, Managers, Designers, Engineers, and Contractors*. Hoboken, NJ: John Wiley & Sons.

ECKBLAD, S., RUBEL, Z. and BEDRICK, J., 2007. *Integrated Project Delivery What, Why and How*. AIA.

ECONOMIST, 2009-last update, **Abu Dhabi's ambitious eco-city** *Masdar plan* [Homepage of The Economist], [Online]. Available: http://www.economist.com/science/tq/displaystory.cfm?story_id=12673433 [15-03-2009].

EGAN, J., 1998. ***Rethinking Construction*** *The Report of the Construction Task Force*. London: Department of the Environment, Transport and the Regions.

ELGOT, J., 2021. **Firm of lobbying inquiry chair won £7m of government contracts in past year**. *The Guardian*, 20 April 2021.

ERHVERVSMINISTERIET, 2019-last update, Digitalisering og produktivitet vaekstpotentiale i danske virksomheder. Available: https://em.dk/aktuelt/udgivelser-og-aftaler/2017/maj/digitalisering-og-produktivitet-vaekstpotentiale-i-danske-virksomheder [16-12-2023].

ERKESSOUSI, N.E., 2010. *How Integrated Project Delivery is an Advantage to the Danish Building Industry, and How It Can Be Executed*. Copenhagen: Copenhagen School of Design & Technology.

ERNSTROM, B., HANSON, D., HILL, D., CLARK, J.J., HOLDER, M.K., TURNER, D.N., SUNDT, D.R., BARTON, L.S.I. and BARTON, T.W., 2010. *The Contractors' Guide to BIM*. Arlington, VA: Association of General Contractors (AGC).

ESSANY, M., 2011-last update, Mobile payments giant square aims to fossilize the cash register [Homepage of Mobile Marketing Watch], [Online]. Available: http://www.mobilemarketingwatch.com/mobile-payments-giant-square-aims-to-fossilize-the-cash-register-15666/ (accessed 2023).

EUROPEAN COMMISSION, 1989-last update, ECTS users' guide - European Commission. Available: http://ec.europa.eu/education/ects/users-guide/introduction_en.htm [4/11/, 2017].

EVANS, D., 2011. ***The Internet of Things*** *How the Next Evolution of the Internet Is Changing Everything*. San Jose, CA: Cisco Internet Business Solutions Group (IBSG).

FARNELL, J., 2018. *How Construction Margins Fail to Make Up for the Risks*. London: Construction Management.

FERROUSSAT, D., 2007. *BA Investor Presentation*. London: BAA.

FERROUSSAT, D., 2008. ***The Terminal 5 Project*** – *Heathrow Airport*. London: BAA Heathrow.

FITCH, D., 2023. *The Blockchain Book | How Blockchain is Enabling The Next Generation*. Gronongen, The Netherlands: BLING.

FLAVELLE, C., 2021-last update, How Trump tried, but largely failed, to Derail America's top climate report [Homepage of New York Times], [Online]. Available: https://www.nytimes.com/2021/01/01/climate/trump-national-climate-assessment.html?action=click&module=Top%20Stories&pgtype=Homepage [02-01-2021].

FLODSTRÖM, A., 1999. Engineering education in the future. In: H.S. OLESEN and J.H. JENSEN, eds, ***Project Studies*** *– A Late Modern University Reform?* Roskilde: Roskilde University Press, pp. 57–66.

FORESTER, J., 11 January, 2018-last update, **BIM and Blockchain: One person's musings on a disruptive technology** [Homepage of AECBytes], [Online]. Available: https://www.aecbytes.com/viewpoint/2018/issue_84.html [03-04-2023].

FORREST, A., 2021. Labour accuses Michael Gove of misleading parliament over £22m PPE contract. *Independent*, 20 August 2021.

FROSTHOLM, M., 2019a-last update, BIM partner. Available: http://byggerietsblockchains.dk/artikel/pionerprojekt-6-bimpartner/ [16-12-2023].

FROSTHOLM, M., 2019b-last update, Build trust. Available: https://byggerietsblockchains-dk.translate.goog/artikel/pionerprojekt-1-buildtrust/?_x_tr_sl=da&_x_tr_tl=en&_x_tr_hl=en&_x_tr_pto=sc&_x_tr_sch=http [16-12-2023].

FRY, S., 24 February 2020-last update, Horses and hunting [Homepage of Subtitles (SRT & ASS) for Movies, Documentaries, +], [Online]. Available: https://subsaga.com/bbc/comedy/qi/series-h/12-horses-and-hunting.html [25-02-2020].

GABRIELLI, J., 2010-last update, **Architecture**: The composite master builder [Homepage of Whole Building Design Guide], [Online]. Available: http://www.wbdg.org/design/dd_architecture.php [09-08-2011].

GALAEV, M., 2018-last update, Why are construction profit-margins so low? [Homepage of KREO], [Online]. Available: https://www.kreo.net/blog/why-are-construction-profit-margins-so-low [16-03-2020].

GEHRY, F., 2008. Digital project – Frank Gehry. [26-11-2008].

GIBBERT, M. and HOEGL, M., 2011. In praise of dissimilarity. *MIT Sloan Management Review*, **52**(4): 20–22.

HARDIN, B., 2009. ***BIM and Construction Management*** *Proven Tools, Methods and Workflows*. Indianapolis, IN: Wiley Publishing.

HARKELL SHORT, J., 2014-last update, The future of cryptocurrency. Available: http://www.harkell.com/TFOC-30-03-14.pdf [28-04-2020].

HARLEY-MCKEOWN, L., 28 April 2021-last update, The Greensill scandal: How finance firm's collapse rocked Westminster [Homepage of Yahoo, Finance], [Online]. Available: https://uk.finance.yahoo.com/news/greensill-capital-scandal-david-cameron-rishi-sunak-lobbying-parliamentary-inquiry-latest-073657007.html?guccounter=1&guce_referrer=aHR0cHM6Ly93d3cuZ29vZ2xlLmNvbS8&guce_referrer_sig=AQAAAK0d-ZumWx8p268FpA8ZEb-6Z8d_RehcZQK43YuGJGCI1d2mwM35wAgbgKWsFGM_RghD0Zw1oyDmQ2NU_6VpORvmeaYqO3_PvL_h5-JLOB5mg3d

iHVVBfwaLGC0my2WdCthJYjSMciYsEe7tbL41mVzQqcWxZAgEjfXVgA9ic ByL [17-08-2021].

HARRIS, T. and RASKIN, A., 4 April 2023-last update, The AI dilemma [Homepage of Center for Humane Technology], [Online]. Available: https://www.youtube.com/watch?v=cUrm9MfKgTs [25-04-2023].

HARTY, J., 2012. *The Impact of Digitalisation on the Management Role of Architectural Technology*. Aberdeen: Robert Gordon University.

HARTY, J. and LAING, R., 2010. In: J. UNDERWOOD and U. ISIKDAG. Chapter 24. Removing barriers to BIM adoption: clients and code checking to drive changes, eds, ***Handbook of Research on Building Information Modeling and Construction Informatics****: Concepts and Technologies*. In: J. UNDERWOOD and U. ISIKDAG, eds First edn. Hersey, NY: Information Science Reference (an imprint of IGI Global), pp. 546–560.

HARTY, J. and LAING, R., 2011. Trust and risk in collaborative environments. In: J. COUNSELL, F. KHOSROWSHAHI and R. LAING, eds, ***Information Visualization****, Visualization of Built & Rural Environments*, 13–15 July 2011. Los Alamitos, CA: Institute of Electrical & Electronic Engineers Computer Society, pp. 558–563.

HARTY, J. and MILLER, R., 2014. **The carbon quota house** – Compared to best practices, it can combine sustainable solutions for better house typologies. In: T. KOUIDER, ed., ***Architectural Technology****, Towards Innovative Professional Practice*, 7 November 2014. Aberdeen: RGU, pp. 49–61.

HASTE, N., 2002. ***Terminal Five Agreement****; The Delivery Team Handbook (without PEP)*. London.

HAWKEN, P., 2017. ***DRAWDOWN*** | *The Most Comprehensive Plan Ever Proposed to Reverse Global Warming*. London: Penguin Books.

HEIKKILÄ, S., 2014. *Mobility as a Service – A Proposal for Action for the Public Administration, Case Helsinki*. Aalto University.

HJELHOLT, M. and SCHOU, J., 2019. Digitalisering af den danske offentlige sektor: hvor er vi på vej hen? *Økonomi & Politik*, **92**(2).

IEA, 2014-last update, **Electric power transmission and distribution losses (% of output) – Country Ranking** [Homepage of International Energy Agency], [Online]. Available: https://www.indexmundi.com/facts/indicators/EG.ELC.LOSS.ZS/rankings [22-12-2020].

ILLERIS, K., 1999. Project work in university studies: Background and current issues. In: H.S. OLESEN and J.H. JENSEN, eds, ***Project Studies*** – *A Late Modern University Reform?* Roskilde: Roskilde University Press, pp. 25–37.

ISIKDAG, U., UNDERWOOD, J., KURUOGLU, M. and ABDUL-RAHMANN, A., 2010. Geospatial views for restful BIM. In: J. UNDERWOOD and U. ISIKDAG, eds, *Handbook of Research on Building Information Modeling and Construction Informatics: Concepts and Technologies*. New York: IGI Global, pp. 473–482.

JARNAGIN, C. and SLOCUM, J.W.J., 2007. Creating corporate cultures through mythopoetic leadership. *Organizational Dynamics*, **36**(3), pp. 288–302.

JERNIGAN, F., 2007. ***BIG BIM*** *Little BIM*. Second edn. Salisbury, MD: 4 Site Press.
JOHANSEN, T.F., 2015. *How Does BIM Contributes to LEAN?* Copenhagen: Copenhagen School of Design & Technology.
KAISER, B., 2019. *Targeted | My Inside Story of CAMBRIDGE ANALYTICA and How TRUMP, BREXIT and FACEBOOK Broke Democracy*. London: Harper Collins Publishers.
KEA, January 2013-last update, Bachelor of Architectural Technology and Construction Management [Homepage of Copenhagen School of Design & Technology], [Online]. Available: http://www.kea.dk/en/programmes/bachelor-degree-ba-programmes/ba-of-architectural-technology-and-construction-management/ [09-03-2013].
KENNEDY, B., 2019-last update, Innovation in Practice – Center Parcs – Sisk [Homepage of Sisk], [Online]. Available: https://vimeo.com/295118963/c174431641 [22-06-2020].
KERSZMAN, G., 1999. Twenty years later. In: H.S. OLESEN and J.H. JENSEN, eds, ***Project Studies*** *– A Late Modern University Reform?* Roskilde: Roskilde University Press, pp. 78–92.
KINNAIRD, C. and GEIPEL, M., 2017. ***Blockchain Technology*** *How the Inventions Behind Bitcoin are Enabling a Network of Trust for the Built Environment*. London: Arups.
KIVINIEMI, A., 2015. ***BIM Education*** *– A Bottleneck in BIM Adoption?* Liverpool: University of Liverpool, School of Arhitecture.
KLEIS, B., 2014. ***The MiniCO2 Houses in Nyborg*** *– Valuable Lessons*. Copenhagen: Realdania.
van der KLINK, M. and BOON, J., 10 Dec, 2002-last update, The investigation of competencies within professional domains [Homepage of Taylor & Francis Online], [Online]. Available: https://www.tandfonline.com/doi/abs/10.1080/13678860110059384 [27-10-2021].
KOOLHAAS, R., 1994. *Delirious New York*. New edn. New York: Moacelli Press.
KOSICKI, M., TSIGKARI, M., TSILIAKOS, M. and ELASHRY, K., 2021. Big data and cloud computing for the built environment. In: M. BOLPAGNI, R. GAVINA and D. RIBERIO, eds, *Industry 4.0 for the Built Environment*. New York: Springer, pp. 131–155.
KRISTENSEN, E.K., 2011. Systemic barriers to a future transformation of the building industry from a buyer controlled to a seller driven industry: an analysis of key systemic variables in the building industry, such as 'procurement model', 'buyer perception', 'production mod', and 'leadership and management', principally in a Danish development context and seen from the perspective of the architect. Robert Gorgon University.
LAING and O'ROURKE, 2014-last update, The Leadenhall Building. London. UK [Homepage of Laing O'Rourke], [Online]. Available: http://www.laingorourke.com/our-projects/all-projects/the-leadenhall-building.aspx [29-03-2020].
LANSKY, J., 2018. Possible state approaches to cryptocurrencies. *Journal of Systems Integration*, 9(1), pp. 19–31.

LATHAM, M., 1994. **Constructing the Team** The final report of the government/industry review of procurement and contractual arrangements in the UK Construction Industry. HMSO.

LAUSTSEN, C., 2021. *Smart Contract Epiphany*. Copenhagen: KEA.

LENDAGER, A. and PÁLMADÓTTIR, A., 9 November 2023-last update, Building with Lava: Mother nature's alternative to concrete comes from volcanos!Available: https://www.youtube.com/watch?v=odBhY1uwUME&t=619s (accessed 2023).

LIAO, C., TENG, Y. and YUAN, S., 2019. Blockchain-based cross-organizational integrated platform for issuing and redeeming reward points. *Proceedings of the Tenth International Symposium on Information and Communication Technology*, Department of Computer Science, National Chiao Tung University, Hsinchu 300, Taiwan, ROC.

LION, R., 2004. *Terminal 5 Single Model Environment - Vision, Reality and Results*. UK: EXCITECH COMPUTERS LIMITED.

LIPTON, E., PROTESS, B. and LEHREN, A.W., 2017. With Trump appointees, a raft of potential conflicts and "no transparancy". *New York Times*.

LITTLEFIELD, D., 2008. World architecture 100. *Building Design*, **2008**, pp. 12.

LOUNSBURY, M. and CRUMLEY, E.T., 2007. New practice creation: An institutional perspective on innovation. *Organization Studies*, **28**(7), pp. 993–1012.

MACINTYRE, D., 2016. **Wapping dispute 30 years on:** How Rupert Murdoch changed labour relations – and newspapers – forever. *The Independent* **News/Media/Press**. Donald Macintyre [21-01-2016].

MACLEAMY, P., 2010a-last update, The future of the building industry (1/5): A tale of three domes – YouTube. Available: http://www.youtube.com/watch?v=4dqW70eoQH4 [03-05-2014].

MACLEAMY, P., 2010b-last update, **The Future of the Building Industry** (3/5): *The Effort Curve* [Homepage of HoK], [Online]. Available: http://www.youtube.com/watch?v=9bUlBYc_Gl4 [10-09-2011].

MACLEAMY, P., 2010c-last update, **The Future of the Building Industry** (5/5): *BIM, BAM, BOOM!* [Homepage of HoK], [Online]. Available: http://www.youtube.com/user/hoknetwork#p/u/38/5IgdcCemevI [10-09-2011].

MAKORTOFF, K., 2021, Cameron and Greensill: A timeline of events in the lobbying scandal. *The Guardian*, 23 April 2021.

MALLESON, A., 2016. Ahead of the game. *RIBA Journal*, **123**(01), pp. 46–47.

MATHEWS, M., 2017-last update, BIM + Blockchain [Homepage of Dublin Institute of Technology], [Online]. Available: https://sites.google.com/site/bimblockchainmalachymathews/ [10-14-2017].

MATHEWS, M., ROBLES, D. and BOWE, B., 2017. **BIM+Blockchain**: A Solution to the Trust Problem in Collaboration? CITA BIM Gathering 2017. 23–24 November 2017, DIT.

MAYNE, T., 2006. ***Change or Perish*** *Report on Integrated Practice*. Washington, DC: AIA.

MAYS, P., Oct, 2014-last update, **Facade Design and Fabrication:** *The Expensive Disconnection* [Homepage of 3D Perspectives – A casual talk about 3D innovation out of the cloud], [Online]. Available: http://perspectives.3ds.com/architecture-engineering-construction/facade-design-and-fabrication-the-expensive-disconnection/ [21-01-2016].

MCAFEE, A. and BRYNJOLFSSON, E., 2008. Investing in the IT that makes a competitive difference. *Harvard Business Review*, (July–August), pp. 99.

METZ, C., 2021. *Genius Makers*. London: Random House Business.

MEZIROW, J., 2009. *Transformative Learning in Action*. San Francisco, CA: John Wiley & Sons.

MILLER, G.H., SUEHIRO, J.M., TOUSCHNER, P.M., CONNOLLY, K.J., MILAN PRICE, B., SMITH, R.P., APARICIO, R. and KELL CORNELL, M., 2008. ***On Compensation*** *Considerations for Teams in a Changing Industry*. Washington, DC: AIA.

MONTAGUE, R., 2023. Smart contracts. *Irish Building*, (1), pp. 70–71.

MOORE-BICK, S.M., 11 June 2019-last update, **Grenfell Tower** *Inquiry* [Homepage of GOV-UK], [Online]. Available: https://www.grenfelltowerinquiry.org.uk/ [11-06-2019].

MORRIS, E., 2007-last update, From Horse Power to Horsepower [Homepage of Access], [Online]. Available: http://www.uctc.net/access/30/Access%20 30%20-%2002%20-%20Horse%20Power.pdf [20-01-2016].

MÜLLER, E., 1997. *The Constructing Architect's Manual*. Horsens: Horsens Polytechnic.

NBS, 2015-last update, **Digital BIM** *toolkit* [Homepage of NBS], [Online]. Available: http://www.thenbs.com/bimtoolkit/bimtoolkit.asp [01-02-2015].

NEIL, A., 2006-last update, **Wapping:** *legacy of Rupert's revolution* [Homepage of The Guardian], [Online]. Available: http://www.theguardian.com/business/2006/jan/15/rupertmurdoch.pressandpublishing [20-01-2016].

NESBITT, N., 2006-last update, smartcodes_part-3 [Homepage of ICC], [Online]. Available: http://media.iccsafe.org/news/misc/smart_codes/smartcodes_part-3.html [11-17-2010].

NESLEN, A., 2019. This article is more than **1 year old Climate change could make insurance too expensive for most people – report**. *The Guardian*. [21-03-2019].

O'CONNOR, P.J., 2009. **Integrated project delivery:** Collaboration through new contract forms, AGC-IPD.

OBONYO, E., 2010. Towards agent-augmented ontologies for educational VDC applications. *Journal of Information Technology in Construction*, **15**, pp. 318.

OECD, 2016-last update, Definition and Selection of Competencies (DeSeCo) [Homepage of OECD], [Online]. Available: https://www.oecd.org/education/skills-beyond-school/definitionandselectionofcompetenciesdeseco.htm [27-10-2021].

OFFSITE MANAGEMENT SCHOOL, 2015-last update, 0:42 / 4:27 **Laing O'Rourke's Leadenhall Building** [Homepage of Laing O'Rourke],

[Online]. Available: https://www.youtube.com/watch?v=BqYCeTxyXmc [29-03-2020].

ØHRSTRØM, B., FREDSLUND, L., FUNCK, A., ERIKSEN, G., MATHIESEN, L., LARSEN, B. and PRÆST, M., 2013. ***Framework for Curriculum*** *The Common and Institutional Sections* ***The Bachelor of Architectural Technology and Construction Management***. Syllabus edn. Denmark: Denmark.

OLESEN, H.S. and JENSEN, J.H., 1999. ***Project Studies*** *– A Late Modern University Reform*. Roskilde: Roskilde University Press.

ONUMA, K.G., 2010. **Location, Location, Location** – *BIM, BIM, BIM. 2010,* **Fall 2010**.

ØRESTED, 5 June 2020-last update, Ørested [Homepage of www.energinet.dk], [Online]. Available: https://privat.orsted.dk/el/kampagner/vindstroem-fra-orsted/2020] (accessed 2023).

OVERLAND, I., 2016. Energy: The missing link in globalization. *Energy Research & Social Science*, **14**, pp. 122–130. Norwegian Institute of International Affairs.

PALSBO, N. and HARTY, J., 2013. ***Quantitative Materials, Dynamic Quantities*** *Material Libraries*. Sheffield: Sheffield Hallam University. International Congress for Architectural Technology.

PAULI, M., 2022. *Towards a Circular Built Environment*. ARUP edn. Copenhagen: Building Green.

PAZLAR, T. and TURK, Z., 2008. **Interoperability in practice**: Geometric data exchange using the IFC standard. *The Electronic Journal of Information Technology in Construction*, **13**, pp. 362–380.

PHILIPS, 13 June 2019-last update, Philips' new intelligent connected lighting system [Homepage of Philips], [Online]. Available: https://youtu.be/1ZYJ4wYGajA [13-06-2019].

POPOVICH, N., ALBECK-RIPKA, L. and PIERRE-LOUIS, K., 2019-last update, 95 environmental rules being rolled back under Trump [Homepage of New York Times], [Online]. Available: https://www.nytimes.com/interactive/2019/climate/trump-environment-rollbacks.html [29-03-2020].

POTTS, K., 2009-last update, **Project management and the changing nature of the quantity surveying profession** – Heathrow Terminal 5 case study [Homepage of Royal Institution of Chartered Surveyors], [Online]. Available: http://www.rics.org/site/scripts/download_info.aspx?fileID=3099&categoryID=564 [31-12-2009].

PRENTICE, D., 2018. Blockchain for the renewable energy industry. University of London, European Energy Centre.

PRESSMAN, A., 2014. *Designing Relationships The Art of Collaboration in Architecture*. New York: Routledge.

RAJKOVIĆ, A., 2023. Relationship between BIM and AI how can we use integration of BIM and AI for project management? Copenhagen: KEA.

RANDALL, T., 23 September 2015-last update, **The Smartest Building in the World** Inside the connected future of architecture [Homepage

of Bloomberg Businessweek], [Online]. Available: https://www.bloomberg.com/features/2015-the-edge-the-worlds-greenest-building/ [12-06-2019].
REGULY, E., 2013. No climate-change deniers to be found in the reinsurance business. *The Globe and Post.*
RIBA, 2007. *Outline Plan of Work.* UK: RIBA.
RIBA, 2013. *Plan of Work.* UK: RIBA.
RICH, E., 1985. *Artificial Intellegence.* Third edn. Singapore: McGraw-Hill.
ROOTH, Ø., 2010. *Norwegian Public Clients and Building SMART.* Copenhagen.
ROSSANT, J. and BAKER, S., 2019. ***Hop, Skip, Go** | How the Transport Revolutioon is Transforming Our Lives.* London: Harper Collins Publishers.
SACK, K. and SCHWARTZ, J., 2018. As storms keep coming, FEMA spends billions in 'cycle' of damage and repair. *New York Times.*
SCHEER, D.R., 2014. ***The Death of Drawing**: Architecture in the Age of Simulation.* New York: Routledge.
SCHLEICHER, A., 2010. The case for 21st-century learning. OECD Home.
SCHÖN, D., 1987. ***Educating the Reflective Practitioner**: Toward a New Design for Teaching and Learning in the Professions.* San Francisco, CA: Proquest Info & Learning.
SCHÖN, D., 1991. *The Reflective Practitioner: How Professionals Think in Action.* London: Ashgate Publishing Ltd.
SCOTT, M., 2006. Wembley costs soar towards £1bn. *The Guardian* **Sport**.
SHELDEN, D.D., 2006. *Building Information Modelling.* Odense: BIPS.
SIGURÐSSON, S.A., 2009. *Benefits of Building Information Modeling.* Copenhagen: Copenhagen School of Design & Technology.
SINCLAIR, D., 2013a. ***Assembling a Collaborative Project Team** Practical Tools Including Multidisciplinary Schedules of Services.* London: RIBA Publishing.
SINCLAIR, D., 2013b. *Guide to Using the **RIBA Plan of Work 2013**.* London: RIBA Publishing.
SLEVIN, D., VIÑES FIESTAS, H., LOVISOLO, S., LATINI, P., KIDNEY, S., DIXSON-DECLEVE, S., PHILIPPONNAT, T., LOEFFLER, K., BROCKMANN, K.L., FABIAN, N., HARTENBERGER, U. and BUKOWSKI, M., 2020. What is the EU Taxonomy? EU Technical Expert Group.
SMITH, D.K. and TARDIF, M., 2009. ***Building Information Modeling** A Strategic Implementation Guide for Architects, Engineers, Constructors and Real Estate Asset Managers.* Hoboken, NJ: John Wiley & Sons Inc.
SMYTH, H. and PRYKE, S., 2006. ***The Management of Complex Projects** A Relationship Approach.* Oxford: Blackwell Publishing.
SMYTH, H. and PRYKE, S., 2008. ***Collaborative Relationships in Construction** Developing Frameworks & Networks.* London: Wiley Blackwell.
SØRENSEN, T., 2010. *BIM in Construction Related Education,* Aalborg University.

STANDARDS AUSTRALIA, 2023. ***Digital Engineering*** *Introducing The Common Data Model*. Sydney: Standards Australia.

STEINER, I.D., 1972. *Group Processes and Group Productivity*. New York: Academic Press.

SWISHER, K., 2019. Owning a car will soon be as quaint as owning a horse. *New York Times*.

TAPSCOTT, D., 27 August 2016a-last update, The blockchain will change EVERYTHING! [Homepage of TED Talks], [Online]. Available: https://www.youtube.com/watch?v=yK6Ldefgbl0 [06-03-2019].

TAPSCOTT, D., 2016b. *Blockchain Revolution*. London: Penguin Random House.

TEICHOLZ, P., 2004. Labor productivity declines in the construction industry: Causes and remedies. *AECBytes*, **Viewpoint 4** (14 April 2004).

THACHER, E.F. and COMPEAU, L.D., 1999. Project-based learning communities at Clarkson University. In: H.S. OLESEN and J.H. JENSEN, eds, ***Project Studies*** *– A Late Modern University Reform?* Roskilde: Roskilde University Press, pp. 25–37.

THOMPSON, S., 2019. The Quality Challenge . *RIBA Journal*, (September/ October), pp. 12.

THROSSELL, D., 2009. **Getting your ducks in a line**: a case study of the Bart's and London Hospital *Design Management and Building Information Modelling* (03 June 2009).

TOBIN, J., 2013-last update, Measuring BIM's disruption: Understanding value networks of BIM/VDC [Homepage of AECBytes], [Online]. Available: http://www.aecbytes.com/feature/2013/BIMdisruption.html [29-04-2020].

TSE, T.K., WONG, K.A. and WONG, K.F., 2008-last update, **The utilisation of building information models in nD modelling**: A study of data interfacing and adoption barriers. Available: http://www.itcon.org/cgi-bin/works/Show?2005_8 [27-11-2008].

Van OOSTROM, C., 28 October 2016-last update, Smart cities: How technology will change our buildings [Homepage of TEDxBerlin], [Online]. Available: https://youtu.be/hT4ZsaZsEgc [12-06-2019].

VENTURI, R., SCOTT BROWN, D. and IZENOUR, S., 1972. *Learning from Las Vegas*. Cambridge, MA: MIT Press.

WADHWA, V. and SALKEVER, A., 2019. *The Driver in the Driverless Car*. Second edn. Oakland, CA: Berrett-Koehler Publishers Inc.

WALKER, M., 2009. **One Market Plaza** Autodesk's Gallery and Offices. In: CHRIS BLYTHE, CHIEF EXECUTIVE, THE CHARTERED INSTITUTE OF BUILDING, ed., *Design Management and Building Information Modelling (BIM)*. Berkshire: CIOB.

WARREN, R., PRICE, J., GRAHAM, E., FORSTENHAEUSLER, N. and VANDERWAL, J., 2018. The projected effect on insects, vertebrates, and plants of limiting global warming to 1.5 degrees C rather than 2 degrees C. *Science*, **360**(6390), pp. 791–795.

WATERHOUSE, R., MORRELL, P., HAMIL, S., COLLARD, S., KING, A., CLARK, N., KELL, A. and KLASCHKA, R., 2011-last update, BIM roundtable discussion [Homepage of NBS], [Online]. Available: http://www.thenbs.com/roundtable/index.asp [10-09-2011].

WEARDEN, G., 2018-last update, Carillion collapse exposed government outsourcing flaws – report [Homepage of The Guardian], [Online]. Available: https://www.theguardian.com/business/2018/jul/09/carillion-collapse-exposed-government-outsourcing-flaws-report [16-03-2020].

WEISS, K.L., ed., 2018. *BLOX*. Copenhagen: Realdania.

WHITEHEAD, A.N., 1929. *The Aims of Education*. New York: The Macmillan Company.

WIKIPEDIA, 2021-last update, Satoshi Nakamoto. Available: https://en.wikipedia.org/wiki/Satoshi_Nakamoto [29-04-2020].

WIKIPEDIA CONTRIBUTORS, 2010-last update, Cloud computing [Homepage of Wikipedia, The Free Encyclopedia], [Online]. Available: http://en.wikipedia.org/w/api.php?action=query&prop=revisions&titles=Cloudcomputing&rvprop=timestamp content&format=xml2010] (accessed 2023).

WILDEMEERSCH, D., 1991. Learning from regularity, irregularity and responsibility. *International Journal of Lifelong Education*, **10**(2), pp. 151–158.

WILLIAMS, C., 2009-last update, Lawyers scared of computers [Homepage of The Register], [Online]. Available: http://www.theregister.co.uk/2009/12/23/cps_paper/ [24-12-2009].

WINCH, G., 2002. ***Managing Construction Projects****: An Information Processing Approach*. Oxford: Blackwell Science.

WORLD RESOURCES INSTITUTE, 26 February 2024-last update, ClimateWatch. Available: climatewatchdata.org.

WYLIE, C., 2019. *Mindfu*k Cambridge Analytica And The Plot To Break America*. First edn. New York: Random Press.

YOUNG, N.W.J., JONES, S.A. and BERNSTEIN, H.M., 2007. ***SmartMarket Report on Interoperability*** *in the Construction Industry*. Bedford, MA: McGraw Hill Construction.

YOUNG, N.W.J., JONES, S.A. and BERNSTEIN, H.M., 2008. ***SmartMarket Report on Building Information Modeling*** – *Transforming Design and Construction to Achieve Greater Industry Production*. Bedford, MA: McGraw Hill Construction.

Index

A

B

C

Transforming the Construction Industry with Blockchain: Enhancing Efficiency, Transparency, and Collaboration, First Edition. James Harty.

J

K

L

M

N

O

P